BIOLOGY TEACHER'S DESK BOOK

BIOLOGY TEACHER'S DESK BOOK

Dorothea Allen

PARKER PUBLISHING COMPANY, INC.
West Nyack, New York

© 1979, by

PARKER PUBLISHING COMPANY, INC.

West Nyack, New York

All rights reserved. No part of this book may be reproduced in any form or by any means, without permission in writing from the publisher.

Library of Congress Cataloging in Publication Data

Allen, Dorothea.
   Biology teacher's desk book.

   Bibliography: p.
   Includes index.
   1. Biology—Study and teaching (Secondary)
I. Title.
QH315.A42   574'.07'11   79-12763
ISBN 0-13-076653-4

Printed in the United States of America

To Mama . . .
    with love and devotion

How This Book Will Help the Biology Teacher

Providing meaningful and practical experiences for students is a challenge that all high school biology teachers face today. The experienced teacher is aware that relevancy must become a reality in today's educational process and that today's students will live a major part of their lives in the 21st century. Our job, then, is to prepare them for a life when, it is projected, people will be more imaginative, have a wider range of interests, be more affected by automation and instrumentation, and have a more profound concern for the ever-increasing importance of environmental limitations in meeting the needs of the world's population. What to teach and how to teach it must be explored critically in terms of indicated trends—increasing student input in the determination of learning programs, greater unification of the learning disciplines, and the individualization of learning rates and programs with built-in options for students. In the achievement of these self-determined goals, today's student needs guidance and direction that only the teacher who remains contemporary can provide.

Planning, organizing, developing, presenting, and evaluating a biology course which is equal to these demands is a physically and intellectually absorbing task, and in meeting the challenge, teachers will find the *Biology Teacher's Desk Book* an invaluable aid. Responding to actual teacher needs, it provides:

- techniques for motivating students to learn more than they had planned
- suggestions for innovative approaches to topics which have *relevance*
- criteria for selection of suitable activities, individualized assignments, and evaluative procedures which emphasize mastery learning and instant feedback for students to assess individual progress being made
- teaching strategies and newer approaches to learning, with specific guidelines for developing autotutorial programs and minicourses
- successful techniques for creating and maintaining high level

interest in topics under consideration, encouraging and guiding the learning process of individual students toward greater independence in this endeavor, and drawing the very best efforts from each student
- effective demonstrations that work for individual or small student groups, with special recommendations for ensuring professionalism in teacher demonstrations
- treatment of interdisciplinary aspects of biology, with practical application of mathematics and chemistry, and with suggestions for interplay with ecology, social studies, and English
- ways of dealing effectively with sensitive and controversial issues such as evolution, population control, genetic engineering, and the use of small animals in research, and ways of encouraging the practice of researching reliable sources of information in support of expressed points of view
- laboratory maintenance procedures for living specimens in the laboratory, preparing stock solutions, and setting standards and priorities for safety in the laboratory
- some formulas, tables, charts, and other pertinent information for teacher reference
- an up-to-date list of selected reading references

This desk book is utilitarian and will serve as an invaluable resource and working reference for high school biology teachers. It is particularly suited to the teacher who is flexibly committed to any one of the currently popular programs (BSCS, Multi-Media, Modular, *et al*), and who wishes to customize the course to the students' needs, interests, abilities, and aspirations. It encourages the use of suggested techniques and approaches in a flexible manner, but always with the learning process of the individual student in clear focus; it advocates being selective via the exploration of new and different techniques, approaches, and methodologies; and it acknowledges the contributions of many successful teachers who have reported their special formulas for achieving an improved teaching-learning situation.

It is only through experimentation with a wide range of strategies and by daring to be innovative that we can discover our greatest potential and develop an effective teaching style; and it is only when that style is recognized as distinctive and professional that it has "class." *Biology Teacher's Desk Book* will help you develop teaching methods that are in the best interests of your students and, in the attainment of your goal, make your efforts a little easier . . . and far more productive.

Dorothea Allen

CONTENTS

How This Book Will Help the Biology Teacher **7**

1 Twenty-five Ways to Increase Student Interest in Biology . 17

Creating an Atmosphere of Biology-in-Action • Building on Natural Curiosity and Student Interests • Importance of Relevance • Student Involvement in Planning, Decision Making, and Evaluating • Contributions Made by Outside Speakers, Students, and Visits • Using Library Resources and General Reading to Supplement the Text • Using Mnemonic Devices to Motivate Learning • Using Games to Stimulate Learning

2 Tested Teaching Techniques That Stimulate Learning 43

Welcoming Students to Biology • Developing a Teaching Style • Achieving Variety and Balance in the Teaching-Learning Process • Maintaining Interest at a High Level • Adjusting Activities to Effect a Change of Pace • Introducing a New Topic • Ways to Channel the Learning Objectives of a Lesson • Individual Involvement in Group Interaction • The Classroom Reference Center • Techniques for Written Reports

3 New and Innovative Teaching Strategies 65

Programmed Learning • Collecting and Assembling Illustrative Materials • Encouraging Students to Assume Responsibility for Their Own Learning • Planning and Conducting Advanced Level Courses • Minicourses • Innovative Practices That Increase Student Involvement and Achievement

4	**Individualized Study Options for Students**	85

Building a Personalized Learning Program • Providing Study Options for Students • Individualizing Learning to Accommodate Ability Levels • Independent Study and Research Projects • Credit for Outside Work • Role of the Open Classroom • Using Contracts

5	**Selecting and Using Effective Demonstrations**	105

Ensuring Professionalism in Teacher Demonstrations • Closed Circuit T.V. for Demonstrations • Demonstrations That Work • Time-Lapse Demonstrations • Demonstrations for Projecting or Taping • Demonstrations for Display • Student Involvement in Demonstrations • Long-Range and On-Going Demonstrations

6	**Unifying the Disciplines—Relating Biology to Total Learning**	121

Mathematics in Biological Studies • Biochemical and Biophysical Patterns • Synchronizing Biology Studies with Other Classes • Socio-Biological Topics • Using Library and Research Materials • Biology and General Education

7	**Social/Moral Issues in the Biology Classroom**	141

Developing Scientific Attitudes • Determination of What Is Life • Social/Moral Issues in Biological Science

8	**Meaningful Use of Instruments and Techniques**	161

Bioinstrumentation • Biochemical Techniques • Multi-Technological Approach

9	**Evaluating Student Progress**	179

Designing an Evaluation Program • Values of Opposing Evaluation Forms • Innovative Evaluation Strategies That Work • Blueprint for Constructing Examinations and Unit Tests • Marking and Grading Test Papers • Encouraging Student Self-Evaluation • Determining Student Grades • Evaluation of the Course

10	**Planning and Maintaining a Laboratory Program**	203

Planning Laboratory Activities • Living Specimens in the Laboratory • Student Involvement in the "Living Laboratory" • Providing New and Innovative Materials and Activities • Safety in the Laboratory • Laboratory Work and Community Relations

Recommended Reading and Reference **221**

Appendices **229**

Index **239**

BIOLOGY TEACHER'S DESK BOOK

1 Twenty-five Ways to Increase Student Interest in Biology

Effective motivation for learning biology can transport high school students into a world of excitement and discovery, where active participation in investigations involving life and the living are accompanied by a high level of personal achievement. This is not a new objective, but neither is it an automatic or built-in feature of a course of study in biology. Actually, it is the fervent hope and dream that every biology teacher holds for his students; and it is not an unattainable goal. To be achieved, it requires a finely attuned interaction in which the teacher supplies the proper catalyst for student reaction—the planning, investigating, probing, questioning, reasoning, understanding, applying, and communicating in productive and rewarding activity that constitute the basics of successful teaching and learning.

High school students' interests are, of course, widespread and varied. In addition to curricular pursuits, they devote much time to TV, sports, extra-curricular events, and other activities which offer strong interest appeal. Somehow, we must arouse sufficient interest in biology so that they will enthusiastically direct a fair share of their time, attention, and energy to this vital study also. To this end, it is important to know and to capitalize on the things that turn students on and to avoid those elements that turn them off.

If, over a period of time, we examine student reactions to a variety of situations, we find:

Things That Turn Students On	Things That Turn Students Off
Studies with relevance	A program that is structured too rigidly
Personal involvement in the determination of the learning program and its goals and objectives	A completely unstructured program
	Stereotyped activities

Variety in activities and learning experiences

Innovative presentations

A clear understanding of the goals and objectives at the outset of a study undertaken

Understanding of benefits to be derived from having attained the objectives set

A clear-cut purpose for each lesson

Opportunities to engage in individualized study

Opportunities to work at individual rates

Situations which are interesting and which present a challenge

Teacher enthusiasm and interest in the course

Teacher interest in student success in the course

Teacher professionalism and resourcefulness

Encouragement to put forth best efforts

Recognition for work accomplished

Flexibility in the program

Lessons that present no challenge

Lack of provision for student input

Work that is too difficult or for which there is inadequate background

Work that is too simple or that smacks of activities experienced in grade school

Programs and lessons which have no clearly defined and/or understood purpose

The atmosphere in which learning takes place can do much to generate interest and industry on the part of students. It should be warm, welcoming, professional, and, above all, permeated by a feeling of *activity*—a feeling that this is where things are really happening, and that the happenings relate to *living* things. As most teachers will agree, this is the type of atmosphere that creates the greatest interest and motivates students to become involved in both planned and self-initiated investigations. By inviting them to participate in purposeful activity, it also encourages the employment of inquiry methods and development of greater independence. Once their interest has become aroused they can absorb something of the spirit of scientific procedure and pursue some personally challenging studies with great satisfaction and enjoyment.

CREATING AN ATMOSPHERE OF BIOLOGY-IN-ACTION

Creating an atmosphere which is conducive to learning through participation relies heavily on student and teacher input and on the general emotional climate as well as the physical features of the learning environment—be it a self-contained, multi-purpose, one-room facility, or an elaborate complex consisting of a separate lecture hall, laboratory, media center, animal room, greenhouse, and outdoor learning site. Whatever the existing facilities, it is *activity* which offers strong motivation. The feeling that things are happening is best developed in a laboratory type of setting that reflects the spirit of cooperative scientific endeavor.

Long-range and on-going student investigative work has motivational value if kept in the open, in clear view. Not only does this generate and sustain interest in what is going on, but it has great potential as a learning experience in its own right and commands respect for the work of others. It attaches a certain quality of importance to the work of the individual or group it represents and this, in turn, contributes significantly to the development of self-confidence and pride in an endeavor which attracts attention. It has been my experience that when students display their work in progress, rather than stashing it away in a closet for "safe keeping," they tend to exhibit a greater involvement and to put forth their best efforts. The recognition which is accorded a job well-done is, after all, a strong motivator. The activity of some students tends to spread and include involvement of others, thus creating an atmosphere of biology-in-action.

It is important, too, that students get the proper impression of the real essence of biological study and the nature of investigative methods by which discoveries are made. In this regard, an advanced student's research project or an on-going research-type teacher study can be invaluable. Students are motivated by any activity that attracts their attention. They show an interest in techniques being employed and in progress being made. If kept relatively simple, some of the routines can be taken over by students who exhibit an interest in trying their hand at the techniques and in assuming the responsibility for their performance. In the process, working along with the teacher, they absorb some of the atmosphere and approach which is directed toward the study of biology; they develop a familiarity with the manipulatory skills and methods employed and become involved in a study which may serve as a springboard for a suggested or related investigation of their own plan and design. This approach has been used with considerable success and can be adapted to a wide variety of topics and personal teacher interests, using some very basic guidelines:

- Relate the study topic to something within the experience of the students and to something to which they can attach a reasonably practical value.
- Plan the study so that it will be professional, serious, interesting, worthwhile, and relevant.
- Use techniques that are basic, simple to master, and applicable to other studies in which students may become involved.
- Plan the study to involve the use of some pieces of equipment.
- Plan the study as a long-range and on-going investigation so that it will spread over a long time period.
- Encourage students to observe in order to learn *what* is being done and *how* and *why* it is being done.
- Answer student questions and accept their suggestions pertaining to the study and/or techniques used.
- Encourage students and coach them in the performance of the techniques and routines being employed.
- Gradually, let those who are interested and equipped take over and be responsible for some of the routines.
- Keep data records available and near the project, either posted on the bulletin board or in a notebook or on a clipboard.
- Examine data records with students and guide them in interpreting the information collected, detecting patterns, and making evaluations.
- Allow the study to be open-ended so that students can plan and carry out extensions and/or variations of the study.
- Be enthusiastic about the research project and equally enthusiastic about student activities, once initiated.

A teacher research project as a continuing study contributes to the activity atmosphere and motivates students to become involved in a cooperative endeavor which investigates, explores, and, hopefully, discovers new information. There is strong evidence that students who have been introduced to investigative procedures via close student-teacher interaction are more successful in the employment of scientific methods in similar studies when undertaken on an individual or small group basis. There are also specific benefits to be enjoyed by both parties:

The Student	**The Teacher**
Develops familiarity with proper use of lab equipment	Extends his knowledge in a study of personal or professional interest
Develops familiarity with laboratory procedures	Is relieved of some of the time-consuming routines as they are taken over by interested students
Practices and perfects some manipulatory skills useful in other investigations	
Develops the ability to collect and analyze data and to formulate conclusions	Reaches many students who respond to this form of motivation with enthusiasm
Builds self-confidence	
Experiences personal satisfaction and a sense of achievement	
Extends knowledge concerning organisms and/or biological materials used	
Views learning as a life-long process	
Is stimulated to become actively involved in biological inquiry	
Absorbs some of the feeling of how a research study is conducted	

There are many topic areas which lend themselves to such ongoing study, but those in which you have a personal or a professional interest should be favored because it is through your own infectious enthusiasm that students are motivated. Samples of some teacher research-type topics are illustrated in Figure 1-1; others, adapted to your personal interests and to the character of your students, will provide stimulation for their involvement.

TEACHER-INITIATED STUDY	NATURE OF RESEARCH-TYPE PROJECT	STUDENT INVOLVEMENT
Plant Lectins for Blood Typing	Search for effective blood typing substances by screening a wide spectrum of plant lectins extracted from seeds	• Collect variety of seeds • Extract lectins from seeds • Perform agglutination test on ABO blood type samples • Maintain blood sample supply • Test and record activity of lectins for identifying blood types • Evaluate results • Use and care for equipment (centrifuge, Waring Blendor) • Follow up study with practical application • Plan further study and/or extensions, using other natural substances
Gene Mutations in Drosophila	Determination of effect of temperature variation on mutation rate in a population of WILD type Drosophila	• Maintain cultures at 20°C, 28°C, or other temperature • Harvest adults, each culture (count by sex, identify and tabulate mutants) • Compute mutation rate, each condition • Evaluate results • Plan testing of other factors (light and chemicals) and other organisms (meal-worms and bacteria)

Fig. 1-1 Student input in on-going, teacher-initiated, research-type activity motivates involvement that may lead to individual investigations

BUILDING ON NATURAL CURIOSITY AND STUDENT INTERESTS

Not unlike other humans, high school students are curious individuals. Also, as in other situations, their curiosity opens the door to knowledge and understanding. It provides strong motivation for becoming involved in learning activities in biology.

Students are naturally attracted to the presence of living organisms. Siamese fighting fish in an aquarium, embryonated hens' eggs being hatched in an incubator, or Hydra somersaulting their way across the flat surface of a culture bowl of pond water—the interest that each arouses encourages observations of the life forms and their life styles.

Living organisms can be used to develop fascinating studies involving structures, functions, adaptations, behavior patterns, genetics, reproduction, development, and interrelationships with other organisms. Each organism possesses some student appeal, and each can provide a proper stimulus for extending a learning experience which begins when the student is first attracted.

Maintaining a Living Laboratory

Maintaining a living laboratory contributes significantly to satisfying natural curiosity about plants and animals, while capitalizing on biology's potential to draw students in. The organisms, if suitably displayed, are excellent attention-getters and lend a dynamic quality to observations and inquiries that cannot be matched by pictures, charts, models, preserved specimens, or other means. In terms of sheer motivational value, they are without equal.

Proper selection and successful maintenance require careful planning to ensure maximum employment of specimens in both formal and informal studies:

- Select organisms on the basis of their ability to attract attention, suitability for being maintained in the classroom, and versatility for use in investigative study that extends the observational experience.
- Encourage students to learn early that proper conditions must be maintained for plant and animal growth.
- Encourage students to consult and be guided by appropriate guidebooks and other reference material for setting up and following regimens for care, maintenance, and study.
- At no time and under no circumstance allow any organism to be intentionally abused or to suffer mistreatment due to neglect or ignorance.
- House organisms appropriately, with clear viewing from as many vantage points as possible.
- Make provisions for organism care during weekends, school holidays, and vacation periods.
- Collect specimens during field trips, making note of the natural habitats for approximate duplication in the indoor setting.
- Do not bring into the laboratory more organisms than are needed or can be properly accommodated.
- Encourage students to contribute appropriate organisms, as available from the vicinity, or from places visited during a holiday or vacation period.

- Plan for a variety of organisms to keep interest at a high level; continue to maintain initial attention-getters while introducing some new ones.
- Do not introduce too many organisms or too many environments at one time; introduce new ones from time to time.
- Order living materials from reputable supply houses and be ready to accommodate them on the specified delivery dates.
- Keep living organisms under observation for awhile prior to the time they are intended for use. This provides time for students to establish an intimate relationship with them and raises the interest level with which they approach studies involving them.

There are many parameters of the *living lab*, each with its own characteristics and specimens uniquely suited for specific studies, but you will also find opportunities for extending the observational experience that is engendered by each. One innovative teacher has reported expanded usage of a simple fresh water aquarium in his classroom by making it available for students to use informally. On a table beside the tank he places a microscope, accessories, and appropriate identification charts, thus enabling his students to sample and examine the water for observation, identification, and classification of its life forms. The same general technique could be applied to other situations involving the inhabitants of soil samples from terraria or other environments; or it could be applied to any interesting creature or collection of specimens. Your selection of appropriate identification manuals for wild flowers, trees, sea shells, insects, birds, and both fresh and salt water organisms will provide students with further motivation as they pursue these fascinating studies.

Turning a Field Study into a Treasure Hunt

Biological treasure hunts, when used to effect a change of pace, can lend excitement and a novel approach to field studies being engaged in by groups of students. They provide spontaneous and infectious motivation for learning and achievement. The concerted team effort motivates students to help each other as they locate, collect, and catalog the required specimens, and the rivalry between teams results in greater accuracy and completeness of the collections. Equally effective when used for review or as an initial learning experience (with proper field guides), treasure hunts can be limited to specific areas of study, such as insects, wild flowers, leaves, or aquatic life forms, or they can be more general in scope, such as the Botanical Treasure Hunt illustrated in Figure 1-2.

BOTANICAL TREASURE HUNT

Within the allotted time, examples of the following specimens are to be found in the assigned area.

Each team will receive 1 point for each specimen collected and properly identified. 20 points will be needed for a perfect score but duplicate specimens will not be honored. (HINT: Organize your team so as to make the best possible use of your time and to avoid the occurrence of duplication and/or omission of any of the specimens listed. Also, to protect the condition of your collected specimens, make good use of the vasculum issued to your team.)

Specimens	Collected	Identified & Tagged	I.D. Verified	Point Score
Angiosperm specimens:				
diffuse root system				
tap root				
adventitious roots				
leaf with parallel venation				
leaf with pinnate venation				
leaf with palmate venation				
herbaceous stem				
woody stem				
monocotyledonous seed				
dicotyledonous seed				
seed with wings				
seed with burrs				
Gymnosperm specimens:				
male strobilus (cone)				
female strobilus (cone)				
winged seed				
Pteridophyte specimens:				
club moss				
frond with sori				
Bryophyte specimens:				
liverwort				
gametophyte moss with sporophyte				
Lichen specimen				

Total = _____

Fig. 1-2 Format for a Botanical Treasure Hunt used to lend excitement via a different approach to field studies

Using Working Models

Most teachers who have used working models and displays in their teaching are enthusiastic about their motivational value. At a workshop for biology teachers representing several suburban high schools, the consensus was that students were grasping concepts more readily and with more enthusiasm to learn them once some of these devices were introduced. Since much of the students' curiosity concerns themselves, they are particularly intrigued by opportunities to discover how their bodies work, and, with some encouragement, some can ingeniously construct their own working models to demonstrate these functions. However, in most cases we must rely on biological supply houses for excellent working models of the DNA molecule, RNA synthesis, and the building of organic molecules; for functional human skeletons and torsos; for functioning heart and lung models; and for flexible knee, shoulder, elbow, and hip joint models.

Students are naturally inquisitive about *how* things work and derive far-reaching benefits from hands-on experiences in which they can:

- manipulate a model of the mechanism involved in the operation of a process or function related to life
- repeat the operation of the model as many times as necessary for a complete understanding of the phenomenon
- reinforce their learning by demonstrating the model to others
- work independently and in accordance with their individual rates of learning

IMPORTANCE OF RELEVANCE

Motivation can be increased if students find relevance in topics studied. If the approaches and activities involved are kept practical in nature, the relevance will be seen more clearly and the learning will be more meaningful. For example, students are more highly motivated to become acquainted with the tools of scientific inquiry when they are introduced via employment in a practical situation to which they can attach some importance.

In the case of the Metric System, used as a laboratory tool, motivation is enhanced when a doorway or wall area is marked with a vertical measuring column calibrated in centimeters. Here, students are welcome to invite their friends before and after school to measure each others' height in metric and gain practical experience with the measuring system they will be using during most of their adult lives. In other

exposures relevant to metric (and therefore relevant to laboratory measuring techniques), you can make provisions for:
- determination of volume (in ml.) of a ½ pt. milk container
- determination of volume (in ml.) of a coke can or bottle
- determination of weight (in g.) of a laboratory mouse
- determination of length (in cm.) of a book, desk, or workbench
- determination of weight (in kg.) of an individual student
- determination of volume (in liters) of an aquarium
- calculation (in cu. cm.) of a student's vital lung capacity

My colleagues and I have found that introducing metric with relevancy helps students to gain a better grasp of the system than is experienced through conversions. They exhibit greater ease in the use of lab equipment, have a better concept of reasonable size and weight ranges when applied to laboratory situations, and develop greater self-confidence, while their understanding of the technique is reinforced. One teacher found that practice sessions measuring colored water in assorted laboratory ware contributed greatly to her students' ability to think metric.

Relevance offers strong stimulation in other cases also, when and where it is clearly defined: population genetics studies based on local surveys or the school population, studies of vital lung capacity of smokers as compared with non-smokers in the class, investigations into possible correlations existing between dietary habits and scholastic achievement as determined by grade averages of honor vs. non-honor students, and the effects of effluence from a strategically located factory on the quantity and quality of life in a nearby waterway. Recycling of organic trash, causes and cures of current and threatening diseases, genetic engineering, population control and longevity, and emphasis on anything that is in vogue are currently "relevant" topics and will undoubtedly yield to other pressing issues in the years to come—when today's students are living as adults in the 21st century. Relevance is keyed to topics relating to the following categories:

- students as individuals
- the quality of life enjoyed by humankind
- the interdependence of life and the environment
- the school and the community
- social and moral issues
- foods, additives, drugs, chemicals, and consumer products of all kinds

The value of studying the currently relevant topics comes from the motivational impact they have for developing the methods of approach to problems. This is the real purpose of biology education—to interest the student in the world in which he lives and to equip him with the scientific methods for interpreting that world. Consequently, it serves as a template for approaching a solution to problems of the future, the problems that will have relevance at that time.

The ability to project thought to future situations can be enhanced by providing opportunities to make predictions based on known information. Using overlay transparencies, projected in part only, encourages students to predict the next stage—the sequence of DNA nucleotides coded by a single replicating strand, the effect of a transplanted irradiated mucleus into an enucleated cell, etc.—of a situation in which conditions have been stated. Using this technique you can help students prepare for the application of methods learned today to problems of the future.

Value sessions also bring relevancy to almost any topic under consideration. Hand-out sheets, distributed in advance of the session, lend an air of importance to the activity and guide the student as he organizes and prepares. A clear statement of the topic helps him to focus on its meaning, and ample space for preparing his brief, from first-hand and lab experiences as well as from reference sources, equips him with a sound basis for making and supporting a value judgment.

STUDENT INVOLVEMENT IN PLANNING, DECISION MAKING, AND EVALUATING

Along with relevance, student input is a key motivating force in planning and conducting dynamic learning experiences. The trend toward a more active part being played by students in making decisions concerning their own education is reflected in the many choices and alternatives that are offered in the curriculum and in courses undertaken. Even within a given course, students often contribute valuable ideas for major topics to be studied, specific investigations to be considered, and the direction that will be taken in an open-ended experiment. Deciding how far to go with a study can be determined to a degree by the class. If sufficient interest is in evidence, studies may be expanded to include in-depth investigations on an individual or small group basis. Where students are provided with options, some may elect certain units of work on a contract basis.

To be successful, student involvement in decision making must not be interpreted to be or allowed to become an unstructured cafeteria-style program. It does, however, recognize the growing independence of students since they share in the responsibility for the direction and success of their own learning.

In tried and tested cases, increased motivation has been found to accompany student involvement in three major areas:

1. *Planning and Preparation:*

   AIM: To allow students to decide their objectives for a given study segment (what they would like to learn about a topic and what they would consider a valuable outcome)

   TECHNIQUES: Guiding individuals and groups in the planning of:

   Field trips and nature walks
   Arrangements for outside speakers
   Displays and demonstrations
   Panel discussion and debates

   Guiding individuals and groups in the preparation of:

   Bibliographies
   Recommended resource materials for media center, library, and classroom
   Tour guides on tape or hand-out sheets for bird walks and nature hikes
   Treasure hunt lists for environment visited
   Itinerary for nature walk

   OUTCOMES OF THE MOTIVATION: Students approach study with greater seriousness of purpose.
   Students assume a more mature attitude toward study.
   Students learn to set goals for what they want to know, and to make plans for how to achieve them.
   Students learn to make value judgments about what is important and relevant.
   Students learn to observe and to interpret what they see.

2. *Decision Making in Modus Operandi:*

   AIM: To guide students in making a wise selection from among the options offered for the manner in which a study will be conducted

TECHNIQUES: Allowing students to make choices after giving due consideration to these criteria:

    Is the choice a good one?
    How will the choice benefit the student?
    Would the same objective be accomplished via an alternative route?

Guiding students in making decisions concerning:

    Which resources and how many to use
    How and where to spend scheduled class/lab time
    Whether to devote additional time in lab, library, or resource center
    Whether to opt for "regular" assignment or an optional one

OUTCOMES OF THE MOTIVATION: Choice of individual carries stronger motivation than decision made by teacher or another.

Students learn to make wiser choices, through practice and an occasional unwise decision.

Students learn to consider and analyze alternatives.

Students learn to identify and make adjustments for individual strengths and weaknesses.

Students learn to recognize that there are more ways than one to satisfy an objective set.

3. *Evaluation:*

    AIM: To help students develop a sense of proportion and awareness by evaluating work done by themselves and by others

    TECHNIQUES: Guiding students in the preparation of evaluation cards for judging investigative work according to objective criteria:

        Is purpose of work clearly stated and easily understood?
        Has work been carefully planned, with attention to background, importance, references, investigative techniques and methods?
        Have conclusions been clearly indicated?
        Is the study too brief or superficial, leaving gaps and unanswered questions?

Has the topic been overly developed?
Has the work been substantiated in authoritative sources?
Does the study have interest value?
What suggestions would you make for improvement?

Guiding students in the preparation of devices for self-evaluation:
Helping one another
Games and puzzles
Written and/or oral reports

OUTCOMES
OF THE
MOTIVATION: Students learn to place emphasis on *learning*.
Students learn to help one another and reinforce their own learning.
Students learn to view quality of their own work: Did they achieve the objectives they set?
Work evaluated by peers motivates students to produce their own best efforts.

CONTRIBUTIONS MADE BY OUTSIDE SPEAKERS, STUDENTS, AND VISITS

Persons other than the teacher can be very influential in motivating students. A different personality is always refreshing and often provides the spark needed to kindle a new interest or the incentive to explore the depths of an established one. Other teachers, biologists in research or industry, and present and former students all have the potential for stimulating your students. When inviting them into your classroom, be sure to capitalize on this potential and on developing it.

Motivation by Outside Speakers

The motivational value of outside speakers comes from the contributions they make to your classroom experience and to the learning program of your students. To ensure this you should follow the SPE plan:

Search out community resources and speakers' bureaus sponsored by colleges, professional associations, and biology-oriented industries. Other teachers of biology or a related science, or a biochemist, geneticist, marine biologist, or conservationist are good prospects. Keeping a card index of those available, with dates scheduled, and provision for recording a "rating" (as determined by his stimulating effect on student learning) for each, will be helpful when scheduling speakers for future occasions.

Prepare the individuals concerned:

1. The speaker should be advised of the student situation (scholarship level, experiences they have had, special interests), topics currently under consideration, and the desired focal point of the presentation. A special request should be made for information concerning employment opportunities in the field 5 to 10 years hence.
2. The students should be properly prepared for the speaker, with background and reasons given for the visit, correlation made to a current study, and advisement of opportunities to ask pertinent questions.

Evaluate the effectiveness of the speaker in terms of the increased enthusiasm for learning after his presentation. I have noted that the motivational impact is often long-range and far-reaching and not always identified in the immediate follow-up discussion with students. Many times a speaker's remarks or demonstrations are cited or applied months later when an individual or group encounters a related topic or similar situation. When finally made, evaluation ratings should be recorded on appropriate index cards for reference.

Encouraging Students to Help One Another

Motivation also results when students learn to help one another. Sharing information gleaned from outside reading, suggesting a special technique or a unique way of doing something, or giving an interim report on how an obstacle has been overcome in a current project all grow out of the experiences of the contributor and, viewed thusly, have a stimulating and rewarding effect. Students who help one another have been found to work harder, just to be in a contributor position, and all find that when learning from one's peers the relative roles they play are often interchanged.

Encourage advanced students to demonstrate their independent studies to those in a first-year class. Not only do they gain in self-confidence and reinforce their own learning by the activity, but the beginners simultaneously learn something about a procedure, concept, or study device designed by another student, or about the level of involvement that students can and do attain. This is *real motivation*, for students find it easy to identify with another who is just one step ahead.

Some advanced students can function in the role of laboratory assistants by assisting first-year students with such manipulatory skills as microscope focusing and viewing, making laboratory measurements and calculations, using equipment, making up chemical solutions of desired concentrations, using chemical indicators, and designing and setting up equipment for experiments and independent studies. Via this approach it is possible to conduct laboratory activities which more closely approximate instruction on a one-to-one basis—as well as giving some relief from the pressures of student-teacher interactions. Care must be taken, however, to avoid any misuse of an advanced student's services; there is no stimulation or incentive for him in routine glass washing or other menial chores.

Visits Offer Stimulation

Visits to research labs or other biology-related facilities are very popular with students. But however delightful a day away from the classroom may be, care must be taken that the trip not turn out to be merely an excursion, a tour through a research laboratory, or a visit for which there has not been adequate preparation by both parties. As with outside speakers, a card file can be maintained for available facilities within traveling distance, and contact maintained with the public relations departments of industries and laboratories that welcome student visitors. While their personnel include very knowledgeable people, they must be advised of the background and interests of the student visitors, and you must take the initiative to ensure the success of the trip. Students must also be properly prepared and specific plans made, especially if there is a possibility that the group will be split up into smaller units during the visitation. For example, a guide sheet distributed to students visiting a microbiological research laboratory might be used to remind them of some highlights:

- Be sure to see the electron microscope.
- Which microorganisms being used in the lab are familiar to you?
- What safety precautions do you see being taken?
- Are any of the microorganisms pathogenic?
- What equipment do you see in use?
- How are experimental data being recorded?
- What problem(s) is/are being investigated?

- Were reference and resource materials available? If so, were they being used?
- Which of the observed procedures have you used in your lab studies?

These sheets, brought back to the classroom for reference in a follow-up discussion, are invaluable in determining to what extent the students have been stimulated to view information gathered in its proper perspective. The extent to which they exhibit a greater incentive to engage in some investigative work employing techniques and procedures demonstrated at the facility visited gives a fair estimate of the motivational value of the visit. A group of teachers using guide sheets for a variety of visitations—to zoos, museums, wild life sanctuaries, game preserves, etc.—have found them to be of great motivational value, especially when there has been student input in their preparation.

USING LIBRARY RESOURCES AND GENERAL READING TO SUPPLEMENT THE TEXT

Much of biology can be presented in far more interesting fashion than in textbooks, and much of its learning occurs in places other than in the classroom/laboratory setting. In introductory courses, where facts may not be as important as the interest developed, general reading and resource materials can have a stimulating effect in awakening interest.

TV specials, popular magazines, monographs, film strips and loops, tapes, and paperbacks (including science fiction) have dynamic appeal because of their dramatic presentation. They can be used to advantage to arouse student interest when introducing a new topic, supplementing text material, or reviewing a unit of work. Available in a wide array of interest and ability levels, there is something for everyone.

Students are attracted to the *now* events or topics in biology, and their interests include all things that relate to life, health, sports, medicine, research procedures, and nature. They should be trained to use *Biological Abstracts*, encouraged to read widely, and guided in the development of their biological literacy concerning these and related topics. The degree of success which each attains can then be measured by the suggestions he makes toward helping a classmate with a difficult concept or situation, his ability to substantiate a claim he makes during a discussion or value session, and the growth in evidence as he progresses from simpler to more professional reading and reference materials. His personal motivation will be in evidence as he becomes actively involved in a discussion or investigation, expresses his depth of understanding of a concept, and as he extends and reinforces his own learning through independent study.

A classroom reading table or reference center places general reading materials close at hand, but the school library or resource center can offer far more complete and effective accommodations. By establishing a close working relationship with its director, you can make this center or the library an important adjunct to your classroom and expose your students to a wealth of stimulating material. This implies a team effort, and how well you perform your role can be determined by your response to the following checklist:

- Have I suggested new titles for addition to those currently available?
- Do I request and use feedback information concerning materials that are the most popular with students?
- Do I take advantage of offers for free and/or examination copies of new materials for consideration?
- Do I enlist the aid of students in appraising and evaluating materials being considered for purchase?
- Do I advise the librarian of topics under consideration in class so that the library will be prepared for student requests?
- Do I enlist the aid of the librarian in making up bibliographies for study topics?
- Do I periodically review present materials and make recommendations for deletion of those which are obsolete?
- Do I constantly search for up-to-date materials and keep current with the general reading?

Keeping general and non-text materials and titles current requires constant up-dating, for changes occur rapidly. Being alert to advertising and catalog listings, and availing yourself of opportunities to obtain examination copies and materials on approval is highly recommended. Biology-oriented magazines, of course, give coverage to current topics, and a selection of subscriptions—*Science World, Science Digest, Scientific American Magazine, Natural History Magazine, Readers' Digest, BioScience, Science Magazine, Audubon Magazine, Sea Frontiers, National Wildlife, National Geographic, Science News Letter,* and *Current Science*—offers a wide spectrum of interests and ability levels to be served while providing opportunities for students to step up their reading from simpler to higher levels.

The news media and TV accounts of a newsworthy item are sometimes overly dramatized, but encourage their use anyway and post newsclippings on the bulletin board or encourage reporting on these items in class discussion periods. The sensationalism with which they are presented serves as an excellent attention-getter and sparks an interest. The details reported can then be researched in authoritative sources to verify their accuracy, and the story can be authenticated or questioned.

Student reading and viewing of general materials should be recorded in some appropriate fashion. For a magazine article read, a listing of title and source may be sufficient, but for a book or a news story researched for its authenticity, a brief report in oral or written form lends greater importance. Motivated to read with a purpose, students read more and express themselves with better understanding—a motivation for continuing their self-education in the future, for evaluating news accounts and general reading for interest value, and for an awareness of the changing nature of biological science brought about by significant breakthroughs.

There are times when you may wish a reading assignment to be limited to a specific topic area. To whet the appetite for this reading, try a dramatic introduction. A well-prepared account of a serendipitous biological breakthrough, a rehearsed reading of an excerpt from the life and/or work of a noted biologist, or an interesting anecdote concerning your personal experience while working with or meeting a contemporary figure in biological research will arouse interest in researching further details via directed reading.

USING MNEMONIC DEVICES TO MOTIVATE LEARNING

There are a few occasions when no natural associations can be made with the things that are a part of student learning; of necessity, these things must be memorized by pure rote. How to remember the order of things occurring in a series, however, can become a fun thing for students and a real motivation for their learning if certain mnemonic devices, with provision for student input in devising the mnemonisms, are introduced. Some that have worked well for my students appear in the following chart:

TOPIC	NAMES LISTED IN ORDER AND SHOWING RELATIONSHIPS	MNEMONISM
Hierarchy of Groupings in Classification Scheme	Kingdom Phylum Class Order Family Genus Species	King Philip Came Over For Graduation Speeches
Phases associated with Mitotic Division	Interphase Prophase Metaphase Anaphase Telophase	I Prepare My Assignments Thoroughly
Sub-stages in Prophase of Meiosis I	Leptotene Zygotene Pachytene Diplotene Diakenesis	Little Zebras Perform Dastardly Deeds

USING GAMES TO STIMULATE LEARNING

Of all the activities you can plan to capture the attention of your students, games are by far the most effective. Students, like other individuals, like to solve problems and puzzles; they find the solution to word games, crossword puzzles, and other "brain teasers" stimulating—and they take pleasure in the learning that is a natural outcome of the experience.

Games Stimulate Learning

While not a new approach to learning, games have only recently become mass produced and available in commercially prepared form. There are board games, simulation games, card games, bingo games, and a variety of word games and puzzles for most major topics in biology. Innovative teachers have also prepared their own games and report that Biology Baseball, Biology Football, Hangman, Spell-Down, Find the Fallacy, Plant Identification, and a Natural Selection simulation game are superb stimulators and welcome diversions from some run-of-the-mill, routine learning experiences. Clearly, we are conscious of the need to employ ingenious methods for presenting fundamentals to our students. Games, if constructed in an imaginative way, provide flexibility throughout the year, and the variety and scope of the games devised provides for maximum motivation.

In general, games should be designed with several thoughts in mind:

- The game should stimulate interest in the topic, biologists, vocabulary, biological laws, theories, and principles that apply.
- The game should help students to review, retain, and apply that which they have learned.
- The game should make learning exciting and interesting.
- The game should be scientifically and biologically accurate and pedagogically sound.
- The game should be entertaining.

Games Can Be Used for Review

Crossword puzzles and word games contribute to the improvement of student word recognition, spelling, and vocabulary building while reviewing a topic studied. With or without a theme, word hunt games and hidden word puzzles are featured in popular student magazines where they attract great attention and seem to explode the myth that high school students lack enthusiasm for learning.

Teacher-made word games and "scrambled words" can be devised for the vocabulary associated with smaller units or topic studies. Personal experience with this technique (see samples, Figure 1-3) verifies its value as an effective motivator, with the sharpening of word recognition, spelling, and association resulting from its employment.

Bingo-type games are also excellent review activities and prove to be highly stimulating to students. The resulting improvement in students' vocabularies and word associations amply justifies the use of MANNO, ZOOLO, HERBO, BIOLOGY BINGO, etc.

Games Can Be Used to Extend Learning

Bingo-type games can also be designed for any major topic, with Embryology, Reproduction, Genetics, Evolution, Microbiology, Natural Selection, Animal & Plant Biology, Nutrition, or Ecosystems & Biomes as central themes. Students find them helpful, not only as review activities, but as a means of extending their knowledge about the topic. For example, when constructing a game of 75 to 80 items for 30 cards of 25 spaces

Unscramble the names of the following substances and find out how they are related:

ESALAMY _ _ O O _ _ _ _
LASATACE _ _ _ _ _ _ _ O
SIPALE _ _ _ _ O _
MYSOLYZE _ _ _ _ O _ _ _
PENSIP _ O _ _ _ O

All of the above substances are _ _ _ _ _ _ _

Unscramble the names of the following animals and find out what they have in common:

VERABE _ O _ O _ O
ROSHE _ _ _ O O
CLAME _ O _ O _
GRITE O _ _ _ _
BARTIB O _ _ O _ O

The above animals are _ _ _ _ _ _ _ _ _ _

Unscramble the names of the following cell structures and find the name of something necessary for life

CLEUSUN O _ _ _ _ O _ _
ENGE O _ _ _
ASPLYMOCT _ O _ _ _ _ _ _ _
MOSOBIRE O _ _ _ _ _ _ _
VELACOU _ _ _ _ _ _ O

All cells require _ _ _ _ _ _ to engage in life processes

Fig. 1-3 Scrambled word games for building vocabulary and word recognition

each for GENETICS BINGO, I have found that the addition of 6 to 8 new terms, not previously studied, adds another dimension to the "review" and extends the students' learning experience—the reinforcement of previous learning is enhanced by its extension.

Games Can Be Used for Evaluation

While many kinds of games are versatile and can be used for evaluation, the ones that encourage a student to concentrate, to apply knowledge, and to solve a hypothetical problem truly test his mastery and understanding of a topic. One I have found to be particularly successful is called *Identify the Impostor* and goes like this:

> A series of 5" x 8" cards is prepared, with each listing the characteristics, size, habitat, and other pertinent information about a plant, animal, molecule, structure, process, or other element or factor of biological importance. A diagram, picture, or illustration may be included, and a classification of the subject given. On the basis of the information listed, the student must verify the classification as correct—or identify it as an impostor and give its proper classification. The correct classification, printed on the reverse side of the cards, enables students to use them for self-evaluation, either individually or in small groups. They enjoy this form of evaluation and test each other, with competition between teams or whole class participation sometimes resulting. Additionally, some students have prepared original cards and descriptions to add to the growing number of *impostors* for evaluation.

Students Can Participate in the Preparation of Board Games

Board games offer opportunities for student participation in the preparation of games which can be consciously and precisely geared to the classroom objectives. Lance E. Bedwell has devised an easy-to-follow procedure for developing environmental education games* that can be readily applied to biological topics. He recommends the use of simple materials and suggests that students can assist with:

- drawing up game boards on manila file folders according to the design and color scheme shown in Figure 1-4
- preparing place markers from bottle caps or other small flat objects, painted to correspond with the colors specified on the board
- obtaining a die or a game spinner for use with each board
- preparing packets of question cards from 3" x 5" file cards, color-coded to the board and markers
- preparing a packet of problem cards with penalties—"skip your

---

*Lance E. Bedwell, "Developing Environmental Education Games," reprinted from *The American Biology Teacher*, Vol. 39, No. 3, March 1977; courtesy of the National Association of Biology Teachers, Inc., 11250 Roger Bacon Drive, Reston, Virginia 22090, and with permission of the author.

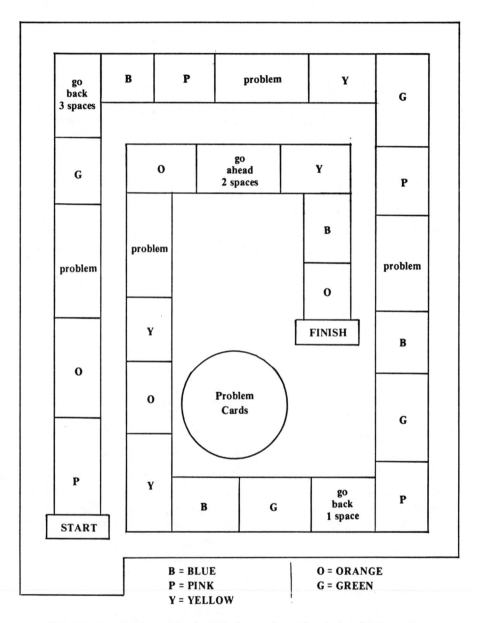

Fig. 1-4 Format for game board, drawn from the design by Lance E. Bedwell for "Developing Environmental Education Games," THE AMERICAN BIOLOGY TEACHER, Vol. 39, No. 3, March 1977; courtesy of The National Association of Biology Teachers, Inc., 11250 Roger Bacon Drive, Reston, Virginia 22090 and with permission of the author

next turn," "player on left loses one turn," "return to the starting point"—to be used when a marker lands on a problem space
- formulating a list of specific rules for playing the game

Games such as *Photosynthesis, Cellular Respiration,* and *Population Genetics* can be developed for use by biology students, using the basic procedure outlined by Bedwell. For the development of board games, he lists the following six basic steps:

Step 1: Develop learner objectives based on the topic studied.

Step 2: Prepare evaluation procedures to ascertain the progress of the students and the effectiveness of the game in achieving the objectives.

Step 3: Write one question, with an appropriate picture or illustration where applicable, on the top card of each packet of question cards.

Step 4: Conduct a test run of the game with a small group of students, and solicit their opinions and ideas for completing the question cards in all packets as you develop the game together.

Step 5: Put the game into use by teams of 5 students (4 players and 1 leader), or by 4 groups of students, with the teacher as the leader who will determine the appropriateness of answers given as each player responds to his/her question card.

Step 6: Evaluate the game to determine areas of strength and weakness; and make revisions, as necessary, to ensure maximum effectiveness in achieving the objectives.

Students do not lack—nor need they lose—interest in biology. But to arouse student interest which will be sustained, a variety of methods must be employed. From the first day of the school term to the last, the enterprising teacher can enlist the aid of motivating and stimulating devices that will increase student involvement in exciting and profitable activities so that they will approach their learning experiences with enthusiasm and come away with the ability to understand and interpret that which is biological science.

2 Tested Teaching Techniques That Stimulate Learning

Introducing high school students to the study of biology is an important prelude to an interesting and productive term in the course; it sets the tone for student experiences and activities for the year and can determine, in some measure, the degree of success that each will achieve. When beginning a new course, students are eager to know what is in store for them; they are concerned with the requirements, the grading system, the kinds of assignments, topics to be studied, and the methods by which they will be studied. But mostly, students today are concerned with the content and objectives of biology, translated in terms of relevancy and benefit to them as individuals. In short, they want some reassurance and justification for having signed up for the course in the first place.

Of course, all biology teachers recognize that these are matters that must be communicated to students, and they employ a variety of methods while doing so. Some find that a *biology student's handbook*, prepared and distributed early in the term, serves admirably while injecting a fresh approach and adding a new dimension to communicating with students. With much room for variation and individualization as befits the teacher, students, and course, the standard inclusions are generally:

- A brief description and topic outline of the course
- Course objectives, both general and specific
- The grading system
- Format for written reports
- Safety practices in the laboratory
- Time distribution for topics studied

It is also helpful if sections devoted to suggestions for correct study habits, guidelines for use of supplementary materials, and procedures for extra assignments, special investigations, and projects are included and made specific so that students can make a direct association between the handbook, the course, and their personal involvement. If you inject a more personal touch as well, such as an explanation of the character of good lab work, encouragement to participate in all activities and open-ended laboratory investigations, and an invitation for students to consult with you on a one-to-one basis, a feeling of interaction between yourself and your students can be established early and developed as a natural follow-up throughout the term. One teacher I know provides for an even greater personalization of the handbook. She includes:

- A calendar for students to mark dates of pertinent events (reports and presentations, field trips, end of marking periods, etc.)
- A student progress report section for each student to record his progress in terms of objectives mastered, as determined by self and teacher evaluations
- A diary for students to log impressions of topics that need further study, those that would be interesting to pursue in greater depth, and questions that are still unanswered.

One advantage here is that when a project or independent study is being planned, each student will have on record his own notations concerning topics with which he has already identified an appreciable interest. As an extra bonus, this very successful teacher finds that personalizing the handbook to this extent makes it such a helpful and valuable reference that students are less likely to lose it and that, with some input of their own, they tend to organize and prepare themselves better in all matters related to the course.

Verbal communications to students should be used to supplement the written form. As new experiences and assignments are encountered, opportunities for discussion and marginal note-taking enhance the value of the handbook as an instructional device by further increasing student input and by adding to the built-in savings in teacher time and energy.

WELCOMING STUDENTS TO BIOLOGY

In my own teaching, I have found a variation of the handbook—*Guidelines for Biology Students*—to be effective. The first installment of a series of printed materials distributed to students is in the form of a "welcome to biology" (see Figure 2-1). Mimeographed on notebook size (8½" x 11") paper, it becomes the first part of the student notebook and includes a general introduction to the course, with guidelines for enjoy-

GUIDELINES FOR BIOLOGY STUDENTS

WELCOME TO BIOLOGY! You have selected an interesting course of study to pursue this year. These guidelines should help you to make the year one that is successful in every way.

BIOLOGY is an **activity** study. You will have many opportunities to observe and investigate living things, the problems they face, and how they solve their problems.

BIOLOGY is an interest-oriented study. You will be able to pursue investigations into topics in which you may have, or in which you may develop an interest.

You are urged to participate in all classroom and laboratory activities. Taking an active part in discussion, reporting on outside reading, suggesting individual and/or group investigations, and preparing a showcase display are but a few of the ways that your interest and awareness can be expressed. Remember, this is YOUR class. Be an active member of it.

You are invited to spend extra time on biological studies or reference work. Feel free to arrange for before-school, after-school, or free periods for working on special assignments.

You are urged to use the library and classroom resources for researching topics under study by the class or special groups, or by individuals. Reporting back to class on your findings can add interest and value to topics for all members of the group.

Throughout the year all matters will be fully explained and you should never be in doubt. You are invited to consult with the teacher for extra help when needed or desired. Don't hesitate to ask for help or to seek additional explanations about an assignment or any other part of your course work.

You should arrange with a class member to be your "Buddy"... to act as a source of information in case of absence from class. The name and telephone number of your buddy should be noted here for use in case of an emergency or an unavoidable absence.

Name of Buddy	Telephone Number

BIOLOGY is an interesting study. Follow all guidelines supplied, work hard, and ENJOY A GOOD YEAR!

Fig. 2-1 WELCOME TO BIOLOGY - first page of Guidelines for Biology Students

able and successful activities. It endeavors to describe the essence of biology in terms of student concerns and to allay some of the students' fears and uncertainties by encouraging them to become acquainted with their classmates and to select a "buddy," by urging them to become actively involved in the exciting experiences of class and lab studies, by advising them to consult with the teacher, and by advising them of what to do in case of an unavoidable absence from a scheduled class.

Supplemental Guidelines

Supplements to the initial pages of *Guidelines* are issued at strategic times throughout the term to accompany assigned or other scheduled activities, or to offer some suggestions for a special situation. Specific sections are devoted to:

- How to Study
- How to Prepare a Research Paper
- How to Write a Lab Report
- How to Plan & Organize Work for Long-Range Assignments
- How to Approach a Lab Session

These are distributed at appropriate times, with discussion and further explanation provided when students have questions or do not understand. Thus, topics described in *Guidelines* are all ones that have also received class time and attention. However, the value of having them in printed form for permanent reference becomes apparent when student attitude and performance reflect positive results, when more and more students appear for informal visits to discuss individual progress, alternate investigations, and extra studies, and when students seem to exhibit a growing ability to organize their work and use reference materials and study guides. It appears that students are genuinely concerned with their own learning programs and that they respond favorably when effort is exerted to help them attain their goals. The ever-present *Guidelines* in the notebook is a tangible reassurance that their success really matters and it serves as a reminder that help is available when needed.

Care must be taken in preparing these written communications, but once a form is completed, it can be varied to suit different class levels or different approaches to the study. Also, it should be examined routinely and critically each year so that additions, deletions, substitutions, and revisions can be made as needed. Content and approach must be constantly up-graded and language and expression up-dated. Its purpose could be easily defeated if a description from last year no longer fits a currently used approach, or if a language gap is permitted to develop between you and your students.

DEVELOPING A TEACHING STYLE

While accomplishing basically the same thing, techniques for communicating information to students may be varied, with many options for teachers. It is surprising when one observes how even a single isolated technique can take on a totally different character as each teacher employs it. Many of us have tried to emulate a favorite college professor or an admired colleague, only to find that the effectiveness of the technique was largely due to the influence of the master's personality, experience, special interests and abilities, or manner of relating with his students.

For most of us, it is necessary to try a technique several times—revising, refining, reorganizing, and re-evaluating it repeatedly before it is accompanied by a feeling of belonging, of comfort, and of satisfaction. This requires an enormous teacher input and, in the process, student feedback can be most helpful. When extended to the sum total of all of the techniques used, its synthesis is inevitable. Out of it evolves the teacher's style which becomes the hallmark by which she or he is identified. It is a rewarding experience, because it is at this point that the *how* becomes at least as important as the *what* of your teaching and its impact and effectiveness on student learning. It is at this point that you begin to sense a feeling that you have arrived as a professional.

Experimenting with different approaches will help you to discover which are consistent with your style in eliciting student effort and achievement beyond the minimum. Keeping individualized learning and learning rates in mind, try the following:

(1) Ask individual students to justify their decisions about a topic under investigation.

(2) Encourage students to increase their depth of study by posing questions such as, "What if just one of the environmental factors in your experiment were changed?" and, "Will this species (being studied) survive the next 100,000 years?"

(3) Present a situation—for example, substances of different solubilities separated by a semi-permeable membrane—via a previously prepared transparency and overhead projector, for students to predict what will happen. Then let them determine how they can check the accuracy of their prediction, while reserving final judgment until all results are available.

(4) Keep on hand, and readily available, copies of *Oxford/Carolina Biology Readers* (available from Carolina Biological Supply Company), *Scientific American* offprints, and *BSCS* pamphlets for on-the-spot distribution to any student indicating an interest in a given topic. Striking while the iron is hot, I find, is more effective than library research as the initial thrust in getting student research under way.

Actually, style is a dynamic entity, ever-changing in response to students and to technological progress. New technologies are constantly being developed, and teachers must be vigilantly resourceful and innovative in the development of new approaches and in the incorporation of new techniques into the teaching-learning situation. Often a new or innovative method is in tune with the "in" things in the student's experience, and consequently it has a more stimulating effect upon his learning—simply because of his ability to identify with it and to relate to it in a positive way.

ACHIEVING VARIETY AND BALANCE IN THE TEACHING-LEARNING PROCESS

The effectiveness of teaching style is also dependent upon the students themselves. Their possible reactions to each situation should be considered when planning a lesson. Some respond best in discussion groups where they can express themselves freely and openly, while others tend to favor a self-directed learning experience in the form of an individual project or laboratory investigation. There are also those for whom a combination of approaches holds the key for maximizing the learning situation. In general, however, high school students are flexible, adjusting readily to a variety of approaches and responding best when the program presented is in good balance.

When strategies are varied, even within the space of a single class period, the change of pace provides the needed impetus to keep students alert and interested. One class I visited illustrates this well: a filmstrip, preceded by a short teacher introduction explaining its purpose and briefly surveying its content, was followed by a few well-prepared questions to provoke thought and motivate a summary review of the film by the students themselves. There was unity and balance in the lesson, with a change of activity for students just often enough to sustain interest at a high level, while maintaining continuity throughout the period.

Planning a balanced lesson is not always easy. While almost every major unit of work lends itself to a variety of strategies from lesson to lesson, the interest appeal of the activities varies widely. Exploring and incorporating elements of some of the newer strategies also proves stimulating because of the fresh approach offered. Some teachers report that using programmed learning texts for a part of a period helps students to concentrate on the topic at hand. Others find them useful for their "settling" effect following a very lively discussion or an interruption or disturbance which interferes with carrying out the planned activities. Regardless of the circumstances, normally somewhat predictable even in the most pupil-centered of classrooms, one principle remains inviolate: the teacher—you—must always assume the responsibility for student achievement. That which must occur spontaneously is more demanding than a planned change of activity in a lesson. Your readiness and resourcefulness will be challenged and tested.

Somewhere between the beginning and the end of the term there should be a planned reversal in the dominance of teacher-directed and student-directed activity, with a balance being achieved somewhere near the middle (see Figure 2-2). By a subtle and almost imperceptible

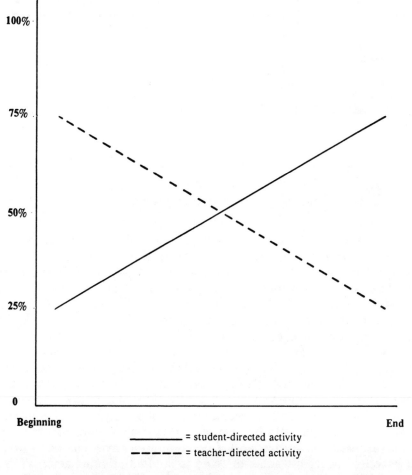

Fig. 2-2 General pattern of reversal in dominance of teacher-directed and student-directed activity in the learning situations occurring from the beginning to the end of a course

shift, the teacher-directed activities should yield to those which give greater emphasis to motivation for learning that is self-directed and for which students exhibit readiness.

MAINTAINING INTEREST AT A HIGH LEVEL

When interest in a topic begins to wane, it can usually be restored to a high level by introducing a new activity that involves all students. For example, when there is a decrease in the number of students participating in a discussion concerning structures and/or functions of some organism studied in the laboratory, a strategic move to a new activity is indicated: writing a short answer to a hypothetical question such as:

What if

... the paramecium didn't have cilia?
... the amoeba couldn't change its shape?
. all mitochondria were removed from muscle cells?

OR

How would you solve your problem if you were

... an embryo of a species whose energy supply is depleted before embryonic development has been completed?
... a euglena cell, recently transferred from a region of light to a region of darkness?

This immediately involves every student. Each becomes involved in an assessment of the problem as he sees it and is encouraged to respond in terms of both immediate and long-range consequences, while focusing attention on the organism, its structure, and its functions, as studied in the laboratory—or on some ingenious way that an organism is equipped to deal with a life and death situation. The change of activity, designed to stimulate new interest in the topic, offers a change of pace from a somewhat mechanical to a more imaginative approach to which all can respond enthusiastically. If you introduce a real attention-getter that involves all students, you can change the activity from one in decline to one that raises the interest level to a new peak.

The change of activity technique is not without possible pitfalls. Lest you get trapped into an unvarying, predictable pattern of changing from discussion to written work, from individual investigation to group interaction, or from teacher-directed to student-oriented activity, it is better that you become attuned to the pulse of the class, anticipating the very moment when the restlessness or boredom begins to set in and averting it quickly by deftly shifting into a new activity with spontaneity and with a built-in change of pace.

In selecting the activities or techniques for inclusion, it is best to be guided by your students—their interests, their ability levels and their span of attention—and by your own assessment of what seems to work best in that all important chemistry that occurs when you and your students react with one another. It is a good idea, too, to experiment with techniques not currently a part of your teaching repertoire. Some worthwhile outcomes result from skillful use of games, puzzles, charades, and other innovative means designed to reinforce learning while maintaining student interest at a high level. Also, you may be unaware that one of your favorite techniques may have grown old, tired, even worn out. So don't hesitate to experiment with different methodologies. It pays to be innovative. And remember that any method (even a good one) tends to become monotonous if over-used.

ADJUSTING ACTIVITIES TO EFFECT A CHANGE OF PACE

It is generally accepted that class attention can be quickly gained by introducing a new activity. This technique can be employed advantageously to stimulate interest at any time throughout the course. I have found that on the first day a class meets, a speedy change over from completed administrative details to an activity that presents *biology* as a *process* not only sparks student interest for the moment but carries over some enthusiasm for learning about life to subsequent sessions. Such activities must be relatively short and simple. For example, student observation of some fresh water jellyfish in several small battery jars or aquaria (one per student group) has been a successful first-day activity for some of my classes. This has served to capture the imagination of students as they watched, fascinated by the rhythmic beauty of the distinctive movement, the upward pumping and graceful downward glide. In a short time they had raised some questions concerning the mechanics involved in this motion and how it relates to the organism's life cycle.

The next day's session invariably centered about a follow-up of student observations, with references to personal experiences with related salt water forms. Discussion of these topics allowed students to participate freely, and each was encouraged to comment and suggest explanations for the motion. When reasons for some of the thoughts were expressed, it was evident that some students had researched the topic (without benefit of a formal assignment), and they were prepared to give reasons backed by supporting evidence from authoritative sources. The natural pattern—viewing, observing, and collecting data, followed by library research to explain the phenomenon—had been established from day one. Repeated and developed further, this pattern flowed naturally into the succeeding sessions on investigative biology which also included some experimental activities.

For the first day change of pace activity from the not-too-exciting administrative detail to a more stimulating observation of living specimens, other organisms such as active Venus Fly Traps, some Siamese Fighting Fish, or some sensitive Mimosa plants are also effective. Or, if time permits, the preparation of a simple experimental set-up—color changes associated with living but not non-living materials in contact with a phenol red indicator solution—may be used. An association of the color change with a *life* process can be established from the mode of action, the chemical change, and the source of the substance responsible for bringing about the change. Other activities also lend themselves to this approach, and if living specimens are not available, teachers can improvise, using whatever materials are handy. For example, each student can be instructed to roll a sheet of paper into the form of a telescope and, with it held in one hand, focus with one eye on a wall clock while the other eye focuses on his free hand, resting open-palmed beside the 'scope. The resulting "hole-in-the-hand" effect will elicit many hypotheses on the part of the students.

In some instances, if the classroom is also the laboratory, and if no living specimens are available, a model or a display can be used as a focal point, or even a reference to a biologically-oriented news item or TV special can spark a lively discussion. In any event, on the first day when students are introduced to the study of biology, be sure to use an approach that will draw them in and try to avoid anything that might turn them off and out.

INTRODUCING A NEW TOPIC

When introducing a topic for study, it is best if the teacher presents it either in person or on tape. The telling of a dramatic anecdote about the work or a significant contribution of Pasteur, Fleming, Darwin, Redi, Mendel, or other biologist often serves to convey to the student some of the excitement of discovery and to arouse their curiosity to the point of wanting to know how the work of these men and others all comes together in expanding man's knowledge and creating an impact on the quality of life. Issues such as pollution, population control, genetic engineering, and chemical food additives have a natural popular appeal and often can be dramatized as introductions to related topics. It is important to plan carefully and rehearse your presentation; tell enough of the story to whet the appetite, raise some important questions in the minds of the students, make it relevant to their lives, and supply references to complete accounts of the "teaser."

Teacher presentation of a new topic also introduces new terms with their proper pronunciation. When the student hears the term correctly, while simultaneously seeing it in print on a projection screen or chalkboard, recognition, spelling, and pronunciation are imprinted at once. If used in context through description or illustration as well, there is an added reinforcement of the learning and understanding of terms. Practice in verbalizing terms also gives students a first-hand familiarity with words not previously a part of their vocabularies and helps in the building of their self-confidence while working with the concepts of a new topic.

In matters of classification and naming of structures and processes, pointing out the appropriateness of the words and the descriptive and precise meaning they bring to the particular situation fosters a more positive approach to their learning. A translation from the original Greek or Latin, or a student analysis of a new word to identify a familiar prefix and/or root word helps to develop an understandable meaning of the term and, in some cases, of words that are already a part of their working vocabularies. Hence, the derivation and descriptive nature of such words as *hermaphrodite* (Hermes, Greek god + Aphrodite, Greek goddess) and *bactericide* (bacteria + killer) illustrate the appropriateness of the words to particular conditions that no other terms can provide.

One of my colleagues finds that her students learn new terms more readily when they are introduced in small groups of related words containing a common root and/or prefix. She has observed that since using this approach there has been a marked improvement in student word recognition and usage and a distinct lessening of questioning the necessity of having to learn biological terminology. A partial listing of her words-in-series includes:

- ingestion, digestion, egestion, indigestion
- intracellular, extracellular, intercellular, non-cellular
- lysis, hemolysis, plasmolysis, autolysis
- hypotonic, hypertonic, isotonic
- oviparous, ovipositor

Exposed to these techniques, students come to appreciate the preciseness and accuracy that biological terms bring to the study; and there is less resistance to learning terms which they initially had anticipated to be difficult and irrelevant. Once introduced, however, the new terms should be practiced often until they become working parts of the students' vocabularies in both oral and written expression. Then, once established, the technique of attacking new terms as encountered should be developed by students as they work toward greater independence in their study.

WAYS TO CHANNEL THE LEARNING OBJECTIVES OF A LESSON

To give meaning and importance to any given lesson, it is necessary that students be made aware of its purpose and objectives and that they be able to set their sights on what it is that they should be learning by the exposure and the experience. "What do we have to learn this for?" and "Is this what you want us to do?" should not be questions in their minds, whether or not expressed. Each should know exactly *what* it is that he is endeavoring, with the teacher's guidance, to learn; and each should know *why* it is important that he learn and understand it.

Often the purpose and development of a lesson can be outlined on a single handout sheet, listing the *agenda* for the day. Some statement of the topic for study, with a review of previous pertinent studies and a brief outline of main headings (with ample space for students to take notes during the proceedings), serves to give direction to the lesson. Serving as a student guide, it helps them to develop an ability for identifying what is important and for recognizing relationships. An evaluation or "windup" in the form of a meaningful application of the information to a new or different situation or a broader concept, and a follow-up assignment which probes into the next topic in concept development should challenge the student to demonstrate his mastery of the topic. If mimeographed on notebook size paper, agenda sheets (see Figure 2-3) can be placed in the student notebook as organized outlines of topics, with notes and some personal interpretation and application—each contributing to the over-all view of the larger concept under consideration.

Distribution of specially prepared guide sheets lends a certain professionalism, a "touch of class," to your teaching. It sharpens student awareness of the importance of every class lesson and of how each fits into the total picture of a significant concept. On the practical side, the inclusion of an evaluation and/or an assignment as well eliminates the necessity of dictating or writing it on the board, both of which are

LESSON GUIDE — ADVANCED BIOLOGY

TOPIC: How can the activities of microorganisms be thwarted?
REVIEW:
   1. Optimum conditions for growth
   2. Rate of growth under optimum conditions
   3. Consequences of removal of any of these conditions

CONSIDER: Agents which thwart microorganisms
   1. Bactericidal
       mode of action
       specific agents
   2. Bacteriostatic
       mode of action
       specific agents, sources
       natural antibiotic producers

EVALUATE: — 18 hrs. ago 4 students prepared petri dishes:

A. sterile nutrient agar at 45°C was heavily inoculated with a pure culture of *Sarcina subflava* and poured into petri dishes. After solidified, the agar surfaces were spotted in well-defined areas with *Bacillus subtilis*.

B. surfaces of sterile nutrient agar plates were swabbed with a broth culture of *Serratia marcescens*. A paper cut-out letter was placed on the agar surface exposed to UV light for 20 minutes. Letters were removed, covers replaced.

Both types of plates were incubated at 37°C for 18-24 hours. One of each is now before you.

Diagram each plate as it now appears and, using your knowledge and understanding of how microorganisms can be thwarted, explain the phenomena and identify the reactants.

A.   B.

ASSIGNMENT:
   1. Research the mode of action of 3 commonly used antibiotics.
   2. Tell under what circumstances you would recommend the use of a bacteriostatic agent, a bactericidal agent.
   3. Why does an antibiotic not have the same effect on the body cells of the patient as it does on the bacterial cells?

Fig. 2-3 Sample of a lesson guide, outlining the AGENDA

time-consuming and leave room for errors. Additionally, they are extremely helpful for student make-up of work missed due to absence.

Similar guide sheets can be prepared for use when viewing a film or film strip. The objectives should be stated clearly so that students know precisely *why* they are viewing, *what* they should learn from the experience, and *how* it relates to the topic under consideration. Each sheet should outline the activity, with some provision for student input, and be designed to guide him in the completion of his objectives.

Organizing data collected as part of an investigative study also requires special skills which students do not automatically possess. Techniques for developing these skills must be devised. Many teachers have found that a work sheet such as that shown in Figure 2-4 is useful for organizing the data collected by students in their beginning studies and for facilitating their concentration on the main thrust of the investigation at hand. After several encounters with teacher-prepared sheets for recording data from both experimental and non-experimental investigations, students tend to establish a data-chart association that enables them to move with relative ease into the phase that places them in the position of designing their own charts for recording data collected in their more advanced and more individualized studies.

INDIVIDUAL INVOLVEMENT IN GROUP INTERACTION

Despite the trend toward individualized learning programs, there are still occasions which call for students to assemble for a discussion of some matter under consideration. Whether it be a small group discussing some aspect of an investigation or experiment, or an entire class engaged in a value session, the primary consideration must, of course, be the needs and special interests of individual students rather than those of the "entire class." As with other strategies designed for the individualized and personalized approach, the discussion should be student-centered. The teacher is a participant who asks questions, probes situations, poses problems, encourages students to participate, and helps them to develop concepts.

To achieve a working group interaction, students need to be encouraged to express themselves and to feel that what they say will be respected. In this regard the message conveyed in "Achieving starts with believing" is a charmer that has helped students more than any negative put-down. So, if at first the contributions made are merely expressions of opinion, they should be accepted, with questions such as "How could you check that out?" or, "Does your experimental data

PHASE FREQUENCY IN MITOSIS

Name _____ Course # _____ Date _____

The purpose of this investigation is to gather information about the five phases of mitosis as they are viewed in a prepared longitudinal section of an onion *(Allium)* root tip. By counting the number of cells appearing in each phase you can compile data from which you should be able to
    1) determine the relative frequency of each phase, and
    2) approximate the duration of each phase, assuming that the entire process (from the beginning of interphase to the end of telophase) takes about 16 hours to complete.

You can perform this investigation with a partner: select a region of the root tip that shows dividing cells well-distributed throughout it; then, as one person works with the microscope and identifies the phases of each cell observed, the other can record this information in the form of a stroke count on the chart below. To avoid confusion or the possibility of counting the same cell more than once, you will find it helpful to follow the orderly arrangement of root tip cells in vertical columns. Begin your count with a column on the left of your field of vision and count the cells from top to bottom before proceeding to the next column to the right. Continue examining columns of cells until sufficient data have been collected for the indicated calculations.

Phase	Number of Cells in Columns						Total	Frequency (%)	Duration
	1	2	3	4	5	6			
Interphase									
Prophase									
Metaphase									
Anaphase									
Telophase									

Fig. 2-4   Student worksheet to be used for recording data collected in an investigative study of phase frequency and duration in mitosis

support that point of view?" These questions hold students somewhat responsible for the statements being made, but without discouragement from making them. If information is not forthcoming from students, additional questions may be posed: "What further information would be helpful for us to have at this point in order to proceed?" and, "Where can this information be obtained?" These usually elicit responses from some members of the group. In time, students will begin to come to the discussion periods better prepared to give reasons for their statements and to use library and other reference materials to research authoritative sources to support any claims they make. Following the lead taken by the teacher, and given the opportunity to do so, students will also begin to use this technique of asking questions of each other, requiring supporting evidence as they interact in the student-student dialogue. In the process, my colleagues and I have found, students develop the ability to consider and evaluate comments made by others and to help one another understand and learn by sharing information. At the same time, each student is reinforcing his own learning and discovering that learning is an important activity.

In group discussions, the role of the teacher is that of a leader who supplies the necessary guidance to keep thought coherent throughout the session. Unquestionably, there are many ways to lead a discussion and there are no set patterns. Additionally, the teacher must exercise resourcefulness in getting a discussion started and in rekindling the spark should interest in the topic begin to wane. But, as with all modern approaches, the discussion, too, is directed toward individual students. It is apparent that the emphasis has shifted away from the older concept of a student-teacher relation, to a student-student interaction which also includes the teacher; the recitation has been replaced by the more productive group interaction and, in achieving this quality in the discussion, there are techniques to be employed and pitfalls to avoid. Some that have proven effective include:

Guidelines for Leading a Group Interaction

DO sit in a circle with the students so that you become a member of the group

start the discussion by asking thought-provoking questions

encourage all students to participate in an interchange of ideas

keep the discussion relevant to the topic being studied

guide the group in a joint attack on the problem posed

encourage students to gather evidence to support claims made

appraise faulty statements and try to get a student to rethink his position

encourage students to evaluate statements made by others

probe a thoughtful statement to extend the thought beyond the point where the student had intended to leave it

allow sufficient time for thoughtful reflection on a problem, a question, a comment, or an evaluation

encourage students to help one another and to share information

convey the idea that learning is important and that every student must fully understand the specific objectives

guide students in concept formation

allow some flexibility, while guiding the direction of the group interaction

DON'T put students down for any statements made or questions asked

become a lecturer

become a listener of recitations

answer your own questions

set yourself apart from student members of the group

grill students or try to get information from them that they do not have

pressure individuals or the group by setting rigid time limits

permit discussion to wander aimlessly or to get "out-of-hand"

THE CLASSROOM REFERENCE CENTER

Students profit immeasurably from supplementary reading and reference work. It is important that the resources in the school library and/or media center be familiar to all students and proficiency in their use be encouraged and developed. Student achievement is noticeably greater when a working relationship is established with the school librarian; instruction in the location and proper use of *Biological Abstracts* and pertinent periodicals, journals, and multi-media materials, followed by guidance on a one-to-one basis, results in better quality student preparation and participation.

To complement the centralized library learning center and the classroom/laboratory learning facilities, it is both helpful and convenient if a classroom library or resource center is set up. The proximity of these supplemental classroom books, magazines, film loops, and cassettes exposes students to a wide array of selected materials representative of different presentations and approaches—and it makes provisions for different interest and reading ability levels. Materials can be assembled from a variety of sources: long-term loan from the school library, complimentary or examination copies from publishers and suppliers, and department requisitioned instructional materials. Appropriate monographs, pamphlets, and programmed learning series unitexts round out the variety of offerings, with something for everyone and with built-in options for each individual.

The facilities for this "researching" need not be elaborate. A carrel or table, with provisions for using and storing viewers, cassettes, and reading materials, and set in a quiet corner of the classroom or laboratory, affords the necessary privacy for the learner in an atmosphere of biological activity. Materials should be related to the topic currently under consideration and should be changed appropriately to keep pace with new topics as they are introduced. Students should be made welcome to use the facilities during their free or unscheduled time periods and to take materials out on an over-night basis.

Signing out over-night materials can be operated in a manner similar to the usual library card system, or, as is more to my liking, the "borrowing" records can be listed on sheets of paper, ruled to accommodate the name and homeroom number of the borrower and the name and number of the item being borrowed. Form sheets can be mimeographed and arranged, by day and date, in a notebook or on a clipboard, with sheets being transferred to a permanent folder when completed each day. Scanning the sheets enables me to determine which materials are the most popular and to spot the names of students who may be selecting materials that are incompatible with their ability levels. On the basis of this, suggestions are made for more challenging reading or more direct presentation, as readiness seems to be indicated by matching materials to individual students. In most cases, however, students appear to seek their own level and, happily, to raise the level of their reading and improve the quality of their performance as their experience with the technique progresses. However, the over-night rule must be strictly adhered to in order to ensure that materials will be available during the course of the school day. In this regard, the homeroom number on the sign-out sheet helps to locate the "forgetters" early in the day.

Single topic books and monographs appear to enjoy great popularity, and some series booklets also lend themselves admirably to this usage. The *Oak Tree Basic Biology in Colour Series, Row-Peterson Unitexts, Prentice-Hall Foundations of Modern Biology Series, LIFE Science Library Series, BSCS Pamphlets, EMI Programmed Learning Booklets*, plus a multitude of titles in *Scientific American* offprints,

Oxford/Carolina Biology Readers, and paperback biology unit books prepared by the American Education Publications Center, can serve as a nucleus about which to build. Used either at the reading table or on an over-night sign-out basis, they are great reinforcers of learning of specific topics.

Student assistance in making selections is very beneficial; their input here adds to the value of the activity. Their preferences run to cassettes, film loops, and film strips, and include some "teaching" or instructional units which demonstrate how to perform specific laboratory techniques. I have noticed that individualized study that includes laboratory investigation has been greatly enhanced through their use.

But supplying and encouraging the use of supplementary reading and resource materials in the classroom should not be an attempt to replace the central library or media center of the school. It should be an adjunct to those learning centers as well as to the biology classroom and laboratory. Its main advantage is that its nearness, with readily available and pertinent materials, offers convenience and guidance to students in an attempt to further enhance the learning success with a confirmation that mastery, learning, and understanding on the part of each student is of prime importance. Though an easement of the mechanics is afforded the student, he is also given some perspective of his part in the learning process, with his responsibilities clearly indicated, but with the assurance that it is important that he learn and that someone is interested in his progress and achievement. It has been noted that often a student will find something in his reading, viewing, or listening that not only relates to a topic under discussion or an investigation under way, but that captures his fancy to the point of motivating him to embark on an independent study which it suggests. The technique, perhaps because of its convenience factor, enjoys strong favor among students, with both immediate and long-range benefits to individuals in evidence.

TECHNIQUES FOR WRITTEN REPORTS

Periodically, students need to be given opportunities to report in written form their inquiries into topics or problems of major biological importance—How does DNA affect the life of a cell? Do mutations affect the operation of a natural selection? Might gametic reproduction have provided an advantage to the primitive heterotroph?—to name a few.

To be a productive experience, an inquiry into topics such as these requires the development of skills, and students do need help in making a proper attack on the problem or situation being investigated. A guide (see Figure 2-5) for reporting the inquiry also helps them to organize hypotheses and findings, and proceeds logically to a conclusion in a format not unlike that of the familiar laboratory report. Students find that it "flows," and through practice in its usage, they come to establish a pattern for reporting their investigative work that is consistent with the scientific method and provides them with an authority to experience the use of inquiry skills in solving problems associated with life and living organisms. A comparison of papers presented at the end of the term with those from an earlier period will reveal a tremendous growth in reporting skills and reward both you and your students with a feeling of satisfaction and accomplishment.

In another approach to written work and reporting individual investigations, the publication of an annual *Student Biology Abstracts* is a great motivator that has considerable practical value. As a companion piece to the full-scale report documenting a major investigation or re-

FORMAT FOR BIOLOGY PAPER

TOPIC: (Does DNA Affect the Life of a Cell?)

HYPOTHESIS: (Using the faculties of your past experience and knowledge, what do you think the answer will be?)

FINDINGS: (Report information from sources that either support or refute your original hypothesis. These findings will have specifics, examples, and logical deductions. This will constitute the MAJOR portion of your paper.)

CONCLUSION: (Prepare a short paragraph—several sentences—using clear, concise wording which tells in summary what you have found through your research.)

CONCLUDED HYPOTHESIS: (Did your findings support your original hypothesis? If they did, tell briefly what mental sources you used to make your hypothesis. If your original hypothesis did not agree with your findings, explain what led you to the error—for example: misconceptions, fallacy, inaccuracies, etc. Be specific.)

APPENDIX: (Relate your concluded hypothesis to something in your own experience. One short paragraph will be necessary.)

*Reminders: Written assignments of this nature should generally be about 250-300 words in length. Sentences should be complete and grammatically correct. Careful thought and clear expression in your own words should characterize your paper.

Fig. 2-5 Sample of student outline for developing a topic paper

search project, a one-page abstract of that work is typed and duplicated in quantity. The abstracts for each year are assembled into a volume and distributed to individuals, the school library, and the classroom reference center. They are useful reference materials for exploring topics that have been studied and summary findings as reported by the investigators. If appropriate, as indicated by the abstract, the complete study report can be taken out of the files for reference in greater depth.

Students should be encouraged to peruse the work of their predecessors in *Student Abstracts* from previous years. Not only might the research findings be valuable to them in their proposed or active investigations, but the depth of study and the quality of reporting comes under scrutiny as they map out their own planned contributions. It is also helpful if you can negotiate an exchange of student publications with other high schools so that more materials become available to your students and their own work will enjoy a wider circulation.

Abstracts are, of course, identified by topic and student contributor, and great importance is attached to the work because of the personal satisfaction that each student derives from it and because of the prestige and permanence associated with published works. In my experience, it is perhaps the single most effective technique for eliciting from students their very best efforts.

The learning that takes place in our classrooms is greatly enhanced by teaching techniques that provide clear guidelines and allow students to concentrate their energies on the main thrust of a study or an investigation. They seem to profit most from techniques that are long-range in design—those that result in the development of useful skills that can be applied to a variety of learning situations and that have built-in provisions for individual growth in responsibility and independence. Biology students are deserving of the very best techniques that we can employ to guide this growth while facilitating their learning.

3 New and Innovative Teaching Strategies

The findings of research study groups investigating the teaching-learning process indicate that students learn not as a group but as individuals, that they learn at different rates, that the learning proceeds gradually and in an orderly and sequential manner from simple to more complex concepts, and that each student responds best when given the opportunity to share in the responsibility for his own learning and the development of his full potential as an individual. These, together with the advent of newer curricula emphasizing the inquiry and investigative approach to the study of biology, dictate the need for more innovative learning programs to replace the authoritative, teacher-centered modes of instruction, now deemed to be inadequate. Provisions for recognizing and accommodating students in this light must be incorporated in newer strategies designed to replace the old.

The quest for alternatives to traditional approaches has not gone unanswered. A variety of programmed learning units, minicourses, autotutorial programs, contractual learning units, open laboratory situations, and individual and small group learning modules have been developed and are being implemented as viable teaching-learning strategies. Some are proving to be effective both in programs of total commitment and when adapted for use in combination with strong and positive elements of a more traditional approach; they are in tune with up-to-date educational objectives and currently popular approaches to learning, and they may be brought into harmony with a teaching style and/or situation that is continually evolving while remaining contemporary.

PROGRAMMED LEARNING

Programmed learning is a process which is based on educationally sound principles. It also offers a workable solution to the problem of how best to provide effective individualized learning experiences for students:

- It emphasizes *mastery learning*, thereby eliminating any possibility of gaps in an individual's understanding of a topic or a concept.
- It replaces the forced pacing normally imposed on students in a group learning situation, with self-pacing in accordance with the rate at which each individual can absorb the information presented.
- It presents and develops a topic according to the pattern of how students learn—in small steps and in a logical and orderly sequence.
- It involves each student in related and meaningful activity by requiring him to make frequent and active responses to the learning stimuli.
- It supplies each student with an immediate feedback to his responses and a reinforcement of his learning.
- It allows the student to repeat the program, in its entirety or in part, as many times as he finds necessary.

Most programmed instruction is autotutorial in design. It is flexible and adaptable and need not be used in the same way at all times. Actually, its motivational value is enhanced by varying its use as well as its design. Teachers have reported successful employment of the technique in introducing a new topic, reviewing a topic already studied, providing extra help for a slow learner, and providing supplementary material for a topic which was developed according to a different approach or strategy. It also solves the problem of how to provide the individualized instruction needed by students who are engaged in open-ended investigations and independent study projects.

Autotutorial (or AT) programs have enormous potential for improving the effectiveness of your teaching. When considering a replacement for an older strategy or technique, or a supplement for a current one, you will want to research the objectives, mechanics, characteristics, and implementation of the available materials. A number of different devices—varying from sophisticated teaching machines with solid-state computer-like capabilities to cassette/film strip or slide presentations and booklets, with or without an audio commentary—are

NEW AND INNOVATIVE TEACHING STRATEGIES

available commercially. "Frames" can be viewed on a monitor, projected on a viewing screen, or read on a printed page; and student responses can be made by typing on a computer's typewriter, or recorded in the form of short answers to direct questions designed to test the student's mastery and understanding.

Program Forms

Perhaps the most important feature of a program design is the form along which it is developed (see Figures 3-1a and 3-1b):

1. The *linear* form places all material relating to the topic in a

Since the building blocks of all organisms are cells it is reasonable to assume that the reproduction of an organism depends upon the reproduction of its _____.	cells
A protozoan characteristically reproduces by dividing its single cell into 2 identical daughter cells. In its reproductive process the cell must be equipped with some mechanism for (changing/keeping constant) _____, from generation to generation, its kind and amount of genetic material.	keeping constant
The genetic material of a cell is located in its chromosomes. To ensure the genetic continuity of its species, every generation of that species must have (the same number/a different number) _____ of chromosomes. If a parent cell has 4 chromosomes we expect that each daughter cell will have _____ chromosome(s).	the same number, 4

Fig. 3-1a   Example of linear programming

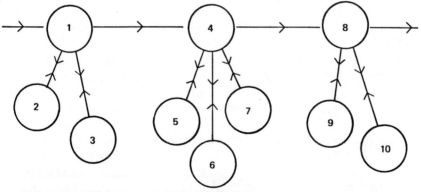

Fig. 3-1b   Example of a simple branched program pattern. On frame 1 the material is followed by three answers of which number 4 is correct. If the student selects answers 2 or 3, he is corrected and directed back to number 1. He is again confronted with multiple choices when he arrives at frames 4 and 8

logical sequence and arranges it from beginning to end for each student to use as he progresses through the entire program.
2. The *branching* form leads a student making an incorrect response to a "branch" program which details or remediates the problem before directing him back to the next step of the main program being followed in an uninterrupted sequence by students who made the correct response.
3. The *predictive* form adapts to students who have some prior knowledge of the topic by using a pretest which predicts which sections of the complete program are suitable for the student's situation and directs his "program" accordingly.

Additional factors to consider are the suitability of available facilities for implementing a program, the use to which the program will be put, and the expected outcomes in terms of an improvement in the teaching-learning process. Some teachers who have had experience with AT programs report that their programs are equally effective when used as an introduction or as a review of a topic, that most programs help students to teach themselves some kinds of factual information, and that the teacher, thus freed from some routine tasks, can utilize the time gained more effectively in conferring with individual students and guiding them in certain aspects of their investigative studies. Teacher-written programs have also been reported as having a highly motivational effect on the students for whom they were specifically written.

How to Write an Autotutorial Program

Writing a "program" is not an activity alien to teachers, regardless of the apparent newness in name. In essence, all have engaged in the activity when preparing a detailed lesson or unit plan for a topic wherein objectives, varied strategies, student activities, reinforcement of learning, and evaluation have been included. Although the writing is time-consuming and demanding, you will find that programs tailored for your own students are those best received. The individuality which characterizes your teaching, plus the personal identification that students make to a customized program, lends a personal quality to which they can relate and, consequently, respond with greater enthusiasm. Hearing a familiar voice—yours—on the taped commentary, viewing the distinctive diagram or other illustrative form on the teaching frames, and responding to questions familiarly phrased, allows for greater identification with the program as an integral part of their total learning experience in the course and gives it an added importance because of the personal touch.

You must, of course, work within the framework of your available materials and, be they computer-type, audio-oriented, or pencil and paper based and incorporated into booklet form, there are basic steps to be followed.

Protocol for Writing a Programmed Learning Unit

1. Develop an outline for the subject matter to be included.
2. Prepare a list of objectives to be met.
3. Arrange the objectives in the sequence most logical for developing the topic.
4. Prepare questions to test student mastery of the subject matter content and attainment of the stated objectives.
5. Prepare the *teaching* frames in fairly large steps, following the linear, branched, or predictive form, as desired.
6. Test-run the program on a student representative of students for whom it was written.
7. On the basis of the results of the test-run, evaluate the program and make necessary revisions.
8. Retest, using the revised form and a different student, and repeat this revision-retesting procedure until you are satisfied with the effectiveness of the resulting program.

To avoid some of the pitfalls that entrap even some of the most experienced programmers, care should be taken in certain areas that bear heavily on the success and effectiveness of a program:

- Limit the subject matter to a few basic concepts that can be clearly developed and supported by factual material and meaningful illustrations and applications.
- Carefully appraise the objectives to ascertain that they are worthwhile and clearly stated, and that they can be reasonably achieved in the program.
- Write the program so that no additional instruction will be needed, but so that it can be integrated with other classroom and laboratory activities.
- Do not use the same programmed instruction for students of widely divergent ability or maturity levels.
- Keep the length of the program within reasonable limits.
- Vary the presentation of teaching and testing frames.

- Vary the number and type of teaching frames in each part of the program.
- Vary the type of question used in the testing frames.
- Do not overprogram; repetition is good but can be overdone.
- Minimize the amount of information passing directly from the program to the student. Instead, develop the program as a process of inquiry which enhances the student's ability to solve problems.
- Endeavor to make a program interesting and challenging by including activities which require application of basic material to new situations and use of thought processes.
- Provide opportunities for students to make active responses: writing out work sheets, answering questions, diagramming a situation, solving a problem, analyzing data, labeling a diagram, or coloring a diagram according to a prescribed color code to help students focus attention on what the program is about.
- Emphasize *mastery learning* by encouraging all students to complete all parts of a program with complete understanding before proceeding to the next step.
- Provide additional, more challenging, and/or more in-depth programs for students engaged in extended and/or independent work.
- Make frames uncluttered and, if to be projected, with type large enough to be viewed clearly.
- Welcome feedback from students; many revisions are the better for having taken into account student comments concerning wording that is awkward or ambiguous, or programs that are too long and tedious or too short and superficial.

COLLECTING AND ASSEMBLING ILLUSTRATIVE MATERIALS

The success of a programmed learning unit depends upon the inclusion of illustrative materials. Not only do these materials enrich the topic investigation, but when positioned strategically in the program development, they also provide a welcome and necessary change of pace. The tiredness and boredom that usually sets in when the student is subjected to a single activity for too long a period of time can thus be averted.

There are many forms of supplementary materials which contribute to the effectiveness of programs covering the diverse topics which make up a total biology program.

Plastic embedded specimens, available from most major biological supply houses, permit small or fragile specimens to be handled by students, with opportunities for clear viewing from all angles.

Living specimens in aquaria, terraria, or other appropriate display, such as an active ant colony or beehive, provide suitable materials for observation of activities associated with organisms as well as comparative studies of their structure and function. Living specimens are most effective when cultured and maintained by students who have collected them during individual or group field work.

Selected, prepared microscope slides, such as Ward's Micro-Explano-Mounts (available from Ward's Natural Science Establishment, Inc.) for both basic botany and zoology, with labeled drawings of the slide subject and an accompanying explanatory text, illustrate pertinent aspects of a program topic.

Pictures, charts, and graphs, clipped from magazines, catalogs, advertising literature and brochures from suppliers, and discarded or examination copies of text and reference books can be laminated or covered with clear plastic or clear mylar protectors to withstand the wear of extensive handling.

Handout sheets, duplicated in quantity from professionally prepared or teacher-made spirit masters, are excellent for program-related topics presented in the form of flow charts, diagrams, and cycles of biological phenomena that students can use as review and study aids.

Experiment kits, such as "Lab-Aids" or teacher-assembled materials, with simple instructions for students, permit individuals or small groups to engage in an independent hands-on laboratory type demonstration of a program-related topic.

Collections of butterflies, shells, insects, wild flowers, and leaves, mounted and assembled by students, provide specimen displays for examination, identification, and classification.

Cartridged 8mm film or single topic film loops allow students to view program-related topics on personal-sized projection equipment.

Reading materials, clipped from newspapers and magazines, or in the form of reprints, offprints, or monographs, and arranged in labeled folders, provide students with pertinent up-to-date supplementary material related to the program topic.

The program should refer the student to the selected illustrative materials for: examining on a self-instruction basis; enrichment; illustrating a particular feature; exposing the student to a related topic; or, sometimes, testing. I have found, for example, that a testing frame which refers students to a display of assorted insect specimens allows them to demonstrate their understanding of the teaching frames on insect structures used for identification and classification; it sharpens their powers of concentration and observation as they apply information from the program with greater enthusiasm than is generally exhibited when responding to a standard testing frame.

When not in active use for a study in progress, materials can be attractively displayed in showcase, shelf, bulletin board, or other appropriate display area for students to examine or to review after having completed a study which involved them.

ENCOURAGING STUDENTS TO ASSUME RESPONSIBILITY FOR THEIR OWN LEARNING

It has been demonstrated that a program designed to keep students totally involved encourages them to develop a greater sense of responsibility for their own mastery of subject matter and an understanding of basic concepts. Further, accompanying this development, there is an evolving independence and self-direction in the approach to learning. Specifically:

1. Allowing students to have a voice in the determination of goals and objectives of a study unit encourages them to work at optimum speed and efficiency toward the attainment of goals which they view as meaningful.
2. Preparing a program that keeps students actively involved—making responses, answering questions, examining illustrative materials, displays, etc.—enhances their powers of concentration, and the favorable results that follow are recognized as outcomes of effective learning patterns being developed.
3. Providing for student self-pacing establishes the best learning rates and patterns for individuals and helps the student to a better self-understanding and assessment to which he makes adjustments.
4. Providing students with optional or free-choice programs motivates them to master the basics and use them as a springboard for investigations into topics in which they express a personal interest.
5. Providing open-ended studies stimulates students to expand

their investigative studies on an individual basis into areas of personal involvement.
6. Arranging small group discussions and value sessions, where emphasis is placed on learning, encourages students to develop a greater responsibility to prepare for the sessions in order that they might make a contribution.
7. Conferring with students individually, regarding some investigation or research project of individual undertaking, stimulates the student to prepare for these informal conferences, during which he can report on progress being made, ask questions, and comment on unusual observations.
8. Maintaining an open classroom/laboratory atmosphere permits diversified activities wherein each student is responsible for pursuing his own work, scheduling his time and activity, and assuming the responsibility for the completion of the objectives set and agreed upon at the beginning of the program.
9. Putting learning on a contractual basis gives students options concerning the kind and amount of work to be accomplished to achieve a given grade, thereby placing the responsibility for fulfilling the contract with the student.

PLANNING AND CONDUCTING ADVANCED LEVEL COURSES

There are many capable and highly motivated students who, having successfully completed a first year in biology, desire to continue with the study. A program of advanced work should be made available to them.

As a follow-up of Biology 1, a second course should build on students' interests and be supported by your enthusiasm. It will require that you tailor the course to your students' expectations and make it exciting; it will make great demands on your time and energy; and it will be a rewarding and satisfying experience. Although the mechanics involved in instituting a new course will vary widely with the school system, and will undoubtedly include administrative and Board of Education recommendation and approval, the success of the course will depend ultimately on the students and their interaction with you during its planning, development, and operation.

The reasons for offering a second course are primarily for college preparation and for enrichment of the high school experience of students who plan to pursue some phase of a biology or biology-related career. College-bound students need to be given opportunities to develop and practice good study habits, to work hard and extend their energies in challenging situations, to become adept at handling abstract and sophisticated biological concepts with ease, to view subject matter in its proper perspective, to become proficient in the performance of manipulative laboratory skills, to understand and appreciate the role of technology and instrumentation in research and in scientific progress, and to have extensive experience in the use of effective skills of expression.

Developing advanced courses to meet these objectives is a long, slow process involving countless revisions of a first draft or working outline until a satisfactory program evolves. Adjustments will need to be made to new situations as they arise and as they relate to the available facilities, the character of the student-teacher involvement and interaction, or to any other factor of the learning environment. However, there are important guidelines to help you develop your course with a minimum of effort and with desired results:

1. Determine, together with students, the desirability of an advanced course and the objectives to be achieved.
2. Determine the most appropriate approach—autotutorial or other—for implementing the program.
3. Prepare an outline for the course content which allows you to vary the emphasis and/or order of topics.
4. Select an appropriate textbook and supplementary materials that are compatible with your teaching style, the approach to be implemented, the nature and level of the subject matter to be included, the stated objectives, and the character of the student group involved.
5. Set up criteria for student eligibility for the course—i.e., high motivation and sincere interest in furthering their biology education, aptitude for lab-oriented investigative study, and high achievement in the prerequisite Biology 1 course as determined by a grade of B or better and recommendation of the teacher.
6. Determine the optimum class size for available space and facilities and adhere to reasonable limits of the optimum number.
7. Schedule the equivalent of three one-hour class periods and four one-hour (two double periods) lab periods per week.

8. Schedule lab periods at the end of the day, if possible, to accommodate completion of lab work that might require more than the scheduled time.
9. Be flexible and revise the program when conditions indicate that this would be desirable.
10. Provide for and require extensive readings in research journals and in supplementary and reference books.
11. Allow students freedom in working and encourage them to develop a sense of responsibility.
12. Keep current and interject new developments in both content areas and methodology into the program.
13. Get feedback of information from college students and edit and revise the course accordingly.
14. Provide opportunities for a learning program which is individualized and personalized.
15. Take a personal interest in the individual project work of students and consult with them on a one-to-one basis concerning progress being made and/or problems encountered.

An Advanced Biology Course

Flexibility within an enrichment program which capitalizes on topics and/or activities observed to have been the most stimulating to students while in Biology 1 will enable advanced students to pursue individual research-type studies concerning topics of personal interest and involvement. For example, a course of study in Advanced Biology developed for second year biology students at the Boonton High School is lab-oriented and centers about the major areas of Cellular Biology, Microbiology, Embryology, Genetics, Physiology, and Behavior, but individual researchers may investigate such specific topics as Factors Affecting Cells Harvested from Cultures Grown In Vitro, Factors Effecting a Sex Reversal in Guppies, and The Mating Preferences of Female Fruitflies. Individual studies are naturally compatible with the broad areas included in the course outline, but they also reflect a personal intrigue or concern which a student may have for a specialized topic suggested by the general outline. While geared to the high school level, the course is more than a continuation of Biology 1; it permits extensive investigation and in-depth study designed to bring about better understanding of basic principles and their application, and it encourages students to become responsive, responsible, and resourceful.

Many techniques employed in the program have contributed to the fulfillment of its objectives:

- Opportunities to engage in research-type investigative studies have been accompanied by greater resourcefulness in both planning and conducting the activities.
- Follow-up of lab work with research of literature and other authoritative sources has resulted in the development of better methods of attack for the determination of a theory or principle involved, or for the determination of the mode of action of an agent involved in an experimental situation.
- Extensive use of resource materials has resulted in more thorough investigation and more complete understanding, with some perspective and appreciation for the subject.
- Use of test questions requiring essay-type answers has brought about a noticeable improvement in written expression skills.
- Exposure to outside resources (hospitals and research labs) has encouraged students to exercise greater care in the performance of lab techniques and has resulted in greater importance being attached to them.
- The use of a complete and professional format for writing up laboratory reports has been accompanied by greater skill in organizing work and greater ability in interpreting data collected.
- Course revisions made as a result of feedback from college students have enabled the course to remain flexible and to continue to offer students the experiences that are most helpful to them in college.
- Occasional 6 a.m. sessions for longer-than-regular school-time lab periods have been invigorating experiences and have helped to keep student interest alive.
- One group of area schools reports involvement of all its advanced biology students in a highly successful Biology Conference. The annual affair is hosted by member schools on a rotating basis, but plans and preparations for each program (see Figure 3-2) are made cooperatively by committees representing all schools. Also, every student attending the conference becomes a member of a group for discussion of a broader aspect of one of the topics presented. Presentation and panel discussion topics reflect the interests and concerns of the students, who often develop the program around a central theme that is both timely

ANNUAL BIOLOGY CONFERENCE PROGRAM

1:00 p.m. **AUDITORIUM** — Welcome and Introductory Remarks

1:10 p.m. **Presentation of Research Papers**
1. Genotype Competition in a *Drosophila* Population Cage
2. Injury Requirements for Initiation of Regeneration in Planaria
3. Chemical vs Non-chemical Pesticides in Roach Control
4. Techniques for Plant and Animal Cell Growth *in vitro*
5. Effects of heavy metal ions on Enzymatic Actions
6. First and Second Division Segregation in *Sordaria fimicola*

2:15 p.m. **CAFETERIA** — Refreshments

2:30 p.m. **MEETING ROOMS** — Panel Discussions
- Room 103 1. "Survival of the Fittest"
- Room 107 2. Tissue and Organ Transplants
- Room 112 3. Endangered Species
- Room 115 4. Cloning of Organisms
- Room 117 5. Enzymes in Medicine — the Anti-metabolite Approach
- Room 118 6. Recombinant DNA Research

3:30 p.m. **AUDITORIUM** — Reports of Panel Chairpersons

4:00 p.m. **Concluding Remarks and Adjournment**

Fig. 3-2 Program prepared by advanced biology students for an annual biology conference held in one of the member schools

and relevant. In all, the conference concept is well received; it engenders a sharing of ideas and information among individuals and encourages a spirit of friendly competition as each strives for excellence in his research work and vies for the honor of having it selected for presentation.

Planning an Advanced Placement Program

Some students can profit from an enrichment program that enables them to engage in the equivalent of a college freshman introductory biology course while attending their own high school. There is a formal curriculum for the purpose of enriched acceleration pointed toward college, and for which college credit can be earned if the course is accompanied by:

1. recognition by a participating college the student plans to enter,
2. high achievement in the course and on the Advanced Placement Biology Examination prepared by the College Entrance Examination Board, and
3. recommendation of the teacher.

In some cases students elect to forego the credit that might be earned but still opt for the Advanced Placement (AP) program because it enables them to gain an advantage when taking Introductory College Biology as a college freshman.

Due to the level of sophistication of the AP course, the College Board offers specific guidelines, with a basic course of study outline surveying the content usually included in a college freshman biology course. The outline is "suggested" only, and allows opportunities for varying the order and areas of emphasis as adjustments are made to its implementation in individual high schools. Students should be carefully screened for admission to the course and, in addition to the eligibility requirements established for other advanced courses, it is recommended that prospective AP students have successfully completed a course in chemistry to equip them more adequately with the necessary background for the biochemical basis of life. Students selected should then be given opportunities for individualized work and extensive exposure to information concerning both factual and conceptual biology. A good college textbook which incorporates the subject matter presented on the suggested outline, suitable resource and reference materials, and an adequately equipped laboratory are essential to the success of a course in Advanced Placement Biology.

There are many versions of AP Biology being taught in high schools today. Teachers have improvised and experimented with techniques and approaches which not only enrich students' experiences but, on the practical side, will help them to score well on the AP Test:

- One teacher collects essay-type questions from previously administered AP tests and uses them as bases for student assignment papers. Using the questions in this way, she helps her

students to gain some knowledge of the breadth and scope of topics upon which they will be expected to expound when they sit to write their official AP tests.
- Another, whose students consistently score high on AP testing, places heavy emphasis on physiological studies, claiming that most of the important biological concepts are incorporated in the foundations and principles of physiology on either the cellular or organismic levels.
- And still another relies heavily on extensive reading in *Scientific American* articles, *Oxford/Carolina Biology Readers*, and *Great Experiments in Biology*, by Gabriel and Fogel in which scientific reports appear in their original form. These readings, together with the college textbook selected on the basis of its inclusion of all topic areas listed on the suggested outline, supply her students with a program which is comprehensive and scholarly.

By whatever techniques employed, high school teachers are finding that AP Biology is one of the most effective and promising established programs for stimulating and motivating the interested, the capable, and the biology-oriented student. Furthermore, the list of high schools and colleges participating in the program is growing—an indication that college-level work completed in high school is meeting with some degree of acceptance and success.

MINICOURSES

Many students have some biology-oriented interests that are limited to a very few specialized topics (e.g., oceanography, space biology, entomology, human sexuality, genetics, and evolution), but neglect to enroll in a full semester course because they lack sufficient interest in the full spectrum of topics usually included. Additionally, some who share these specialized interests may find that the regular course which they have completed did not provide for in-depth study of that particular topic. Unless these interests are identified and accommodated, they may never be developed.

Usually, some topics in a general biology course engender a keener interest than do others, and students are quick to make known their preferences, as well as are teachers to read the signs. Topics with student appeal are also made evident in course selection surveys of the school population at large or in the form of student planning on special days when students are given the opportunity to decide what topics they would like to learn about. Often the amount of interest in a single biology topic, however identified, is sufficient to warrant a limited course—a minicourse—in that specialty.

Starting a Minicourse

In some school systems, requests for new courses are handled through an Instructional Council and, if student demand and teacher readiness are matched with council enthusiasm and administrative recommendation, Board of Education approval is granted and plans are made for the preparation of a course of study.

Although they may vary in length of time, it is generally advisable to limit minicourses to a 4-6 week period or to gear them to the marking period, quarter, or other established time interval operating in your school. As an alternative, often found to be very effective in field or laboratory-oriented studies, concentrated sessions on 4 to 6 successive Saturday mornings are highly successful, if scheduling matters can be resolved for their accommodation.

Minicourses are unique and may represent a part of the biology curriculum that is not necessarily college-oriented. Each course is a brief but coherent study of a single topic; it is self-contained and requires no other form of instruction. The direction and depth of the study is largely determined by the individual student's interest, aptitude, and ability. Thus, students sharing a common interest can pursue its study in widely divergent patterns while focusing on the main thrust of a unified short course addressed to the biology specialty.

Most biology topics, if not a part of a developmental program requiring prerequisites, can be developed in a minicourse format with an AT approach that includes the use of audiotapes, visuals, examination materials, laboratory investigations, and library or media center research and reference materials. As in other AT programs, the teacher is primarily a resource person who devotes much time to conferring with individual students. If lab-oriented, instruction and practice in lab techniques must also be included in order that their employment complement the factual and conceptual developments. For example, an 8-week lab-oriented minicourse in microbiology planned for highly motivated capable students must include instructional film loops and descriptive and demonstrative instruction on microbiological techniques which are necessary for engaging in laboratory activities involved in the major topic areas of the course.

Minicourse in Microbiology (8 Weeks)

1. Using the Microscope for Viewing Microorganisms
2. Relative Size of Bacteria and Other Cells
3. Motility of Bacteria vs. Brownian Motion

NEW AND INNOVATIVE TEACHING STRATEGIES

4. Identification of Bacterial Forms and Groupings
5. Determination of Bacterial Differences
6. Pure and Mixed Cultures
7. Widespread Distribution of Microorganisms
8. Patterns of Population Growth
9. Thwarting the Bacteria and Fungi
10. Significance of the Microorganisms

Minicourses in a Full Semester Course

The BSCS (Biological Sciences Curriculum Study) Minicourse Development Project has created a comprehensive AT Program consisting of 76 minicourses arranged in 12 clusters which can be used individually, or from which individual programs of studies can be mapped out for a full semester of study. While there is no limit placed on the number of clusters that can be accommodated in each semester, students do have the freedom to work at their own rate and can be expected to complete a minimum of 4 to 6 clusters per semester. The flexibility of the clusters allows for tailoring a full program for an individual student, with prescribed required minicourse titles as well as those which might be offered as free choice within a group and those which are optional, for enrichment.

Both students and teachers who have experienced minicourses are enthusiastic about their use:

Benefits to the Student	Benefits to the Teacher
Study of a topic of interest can be pursued without commitment to other topics usually included in a general course.	The teacher is a resource person with time to assist individual students on a one-to-one basis.
The teacher can be consulted when information sought "fits" the program of study or when it is needed in order to proceed.	There is less need for devising and using motivational techniques because students who select a specialized biological study tend to be highly motivated on a self-initiated basis.
Personal choice of a study topic encourages working harder and learning more in a concentrated course which focuses on success for all students.	Mastery learning minimizes the occurrence of failures in the course.
Self-pacing (if program is AT	Exposure to one specialized biology course may lead to interest

oriented) is appealing and without pressures to maintain a rigid schedule.

Interests may grow beyond the topic of professed involvement.

in another, thus adding to the influence and prestige of the department.

The student approach is inclined to be positive, with less need for teacher prodding.

INNOVATIVE PRACTICES THAT INCREASE STUDENT INVOLVEMENT AND ACHIEVEMENT

To be effective, it is essential that we also be innovative. A fresh approach renews enthusiasm and charges course offerings with new life and vigor; through the use of new and different strategies you can help to keep interest in your course alive and increase the involvement and level of achievement of your students.

Many innovative practices employed in specific situations have produced results that have contributed to improved learning situations:

- Appropriate paper back books, including some science fiction titles, have contributed to the development and understanding of important biological concepts.
- A modified form of team teaching, in which a fellow teacher has been asked to serve as a guest lecturer or as a specialized topic resource person, has generated more interest and depth to a topic which might otherwise have been more limited in scope.
- The use of a coloring book to accompany studies in anatomy has resulted in greater student understanding of relationships existing between the organs in organismic systems.
- Setting up a culture bank for accommodating other classes and other schools has encouraged the development of more precise lab skills and a better understanding of conditions which are necessary for growing and maintaining cultures of living microorganisms and invertebrates.
- Setting up a *Biology Book of Records* (à la Guinness) has created greater student awareness of living things and an active interest in learning the reasons for the occurrence of unusual specimens and/or situations.
- The use of closed TV for teaching several classes meeting simultaneously has been observed to have been accompanied by closer attention being paid to the presentation.

- Involvement in an intensified back-to-basics program enabled students to raise their test scores significantly in the area of retention of factual information.
- Focusing attention on students in a Biological Science Forum was reported to have resulted in total involvement of all students and to have narrowed the gap between their subject matter content learning and its applications.

Today's students are uninspired by routine activities such as the mere assemblage of a collection of leaves or insects; instead, they tend to perform best in participation-learning situations that focus on the individual and provide for social interaction with the teacher and other students with whom they share a common interest. The guidance and encouragement they receive in the supportive environment of the small group results in students learning from each other, and the probability that effective learning will take place is therefore significantly increased.

We see that the exciting studies of life need to be complemented by equally exciting and effective strategies that place a high priority on competence and productivity, encourage resourcefulness, develop confidence, and permit independence within a flexible, yet structured, program. Charged with the responsibility of acquiring a full repertoire of such strategies, as well as the development of new ones, the innovative teacher works to ensure that the enthusiasm for learning experienced by students while in the lower grades does not become diminished in his high school biology course—rather, the teacher's goal is to highlight excitement and elevate enthusiasm to new heights via the implementation of the best methods and innovative practices that can be mustered.

4 Individualized Study Options for Students

Comparison studies of learning programs indicate that students acquire information via autotutorial methods as well as or better than by traditional methods, and with the added benefit of self-pacing controlled by the individual as he concentrates on *mastery* of the subject matter content.

But, while autotutorial (or AT) methods help to solve the problem of individualized *rates* of learning, we must recognize that a program dedicated to individualized learning must also address itself to other significant aspects of differences among students. There are differences in the degree to which students are capable of performing the reading and paper-and-pencil tasks demanded by academic work, differences in maturity levels for handling abstract concepts, differences in interests and levels of interest in biological topics, and differences in both the immediate and follow-up goals set by individual students. Provisions must be made in the form of self-taught, one-to-one, small group interactions, or entire class learning experiences to accommodate the unique character of the individual student in each particular situation. However, while some will be a natural outgrowth of individual expression and application, most will require careful planning of a program that is tailored to fit each student.

Individualized learning does not imply, necessarily, a different program for each individual. Neither does it imply that students will decide what they will or will not study at any particular point in time—nor that they alone will determine the content and approach to be used in a given course of study. What it does mean is that the learning program must be student-centered: it must be flexible and provide variations in content, level, and approach to investigative studies; it must allow for student input in the planning and implementation of the studies in which they will engage; it must offer students choices and options in accordance with their individual interests, aptitudes, and desires; and it must afford each student a unique set of conditions that, for him, are optimum and will assure his mastery of important biological information.

An individualized learning program is concerned with the learning process in all of its phases. Consequently, it is dependent on the development of a model which incorporates:

- INDIVIDUALIZATION in the MOTIVATION PHASE
- INDIVIDUALIZATION in the PLANNING PHASE
- INDIVIDUALIZATION in the SUBJECT MATTER CONTENT PHASE
- INDIVIDUALIZATION in the LEARNING ACTIVITIES PHASE
- INDIVIDUALIZATION in the REINFORCEMENT OF LEARNING PHASE
- INDIVIDUALIZATION in the EVALUATION PHASE

It is through individualization of learning that the student's personal interest in biology is accelerated and his capabilities are more fully realized.

BUILDING A PERSONALIZED LEARNING PROGRAM

The motivation accompanying a personalized study enhances both the importance of the topic for the individual student and the value of his "personalized" learning experience. Even though beginning students have not yet had enough exposure and experience to identify all or even most of the areas that might prove to be interesting and exciting studies, the exploratory nature of a first-year course lends itself to building on their personal interests. Consequently, the identified interests and self-initiated investigations of individual students are important considerations for the development of individualized learning programs.

Students have many interests, many of which are biologically-oriented. Most have a pet or have visited a wild animal preserve; have observed an animal engaging in some interesting activity or have themselves been active in some sporting event; have a baby brother or sister at home or have been employed as a babysitter; have built or maintained a terrarium or aquarium or have helped to care for a home garden; have hunted, trapped, or fished or have helped with home canning or other food preparation; have fed birds in winter or have watched their migratory flights or nesting habits; or have made insect, butterfly, or shell collections or have gone on camping trips. In addition, they are aware of themselves as individuals as well as of other family members, friends, and classmates of the same and opposite sex. It is to these experiences, involving some relationship to living things, and therefore to biology, that they can relate and attach some real meaning and importance.

Developing a Scenario

Around each of these interests can be built a personalized *scenario* that develops biological principles and concepts along a unique pathway, instead of a stereotyped one designed for the "average" student. The approach to such a study must be structured in a highly personalized form. For example, setting a student free to learn "all about" some insect in which he has expressed an interest offers no structure to his investigation and may cause him to flounder, whereas developing a scenario, in which each stage of his investigation serves as a springboard for proceeding to another level, opens new vistas to be explored and allows him to progress naturally from that which is familiar to a new learning and discovery experience.

Students should be encouraged to present their personal interests for study and consideration; specimens collected from the environs or expressions of concern made in relation to themselves, to a specific life form, or to life in general can offer strong motivation. Each can be developed along the lines of a scenario, wherein the materials employed can be displayed to attract the attention of other students and involve them in the investigation or in a related one. It is a classical example of how to develop a personal interest into a challenging study in which the student is motivated to learn more than he had planned to, and in which he comes to the realization that each interest can become a dynamic study with many facets. In the process, it also introduces him to scientific methods and their employment in a real situation rather than one contrived in a preplanned learning exercise.

The success experienced at each stage of a scenario (see Figure 4-1) serves as a motivator for proceeding to the next level, and each time an objective is set for a new stage it must be carried to completion, until interest in the topic has been satisfied. However, each scenario must

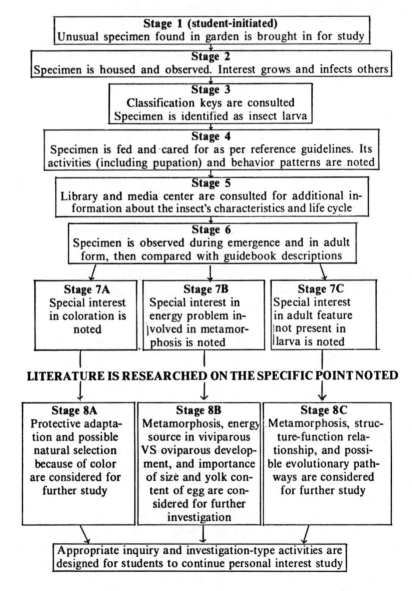

Fig. 4-1 Development of a scenario based on a personal interest

remain flexible so that it can be terminated gracefully at the end of any completed stage, or extended to the next, in accordance with the student's application. This latter determination should be made solely on the basis of the individual student's expressed and continued interest or personal satisfaction with his latest findings. Using a scenario for pursuing student interests results in a greater awareness and better understanding of life and of living things; collected facts are replaced by true understandings in a particular context; and the high level interest initially expressed by an individual is sustained for a longer period of time.

While the flexibility of the scenario might give an impression that the student has an excess of freedom or that he is pursuing only selected aspects of a study of his own choosing, the motivation and planning phases are strong and offer encouragement for the development of the companion phases to their fullest. Thus a student can engage in a highly personalized investigation and, simultaneously, gain the necessary basics for further study, discover with understanding some important biological principles, and develop his ability to make intelligent choices and meaningful application of biological information.

Personalizing Assignments

Students are highly motivated by studies which they can view with relevance and to which they can relate in a personal way. Primarily, these concerns are with themselves and with factors that affect the quality and quantity of life on earth. Pursuing personalized assignments that will strengthen this interest with understanding will also serve to enhance the student's individualized learning program and to sharpen his personal involvement in biological investigative activities. Some personal interest assignment topics that have been used with success include:

- The *now* and the *later* effects of smoking
- The effects of fatigue on reaction time
- The anatomy of a blind spot
- Tracing *my* roots
- Effects of stimulants and depressants on heart rate
- Cardiovascular response to diving
- Radiation injury and tissue metabolism
- Food additives and health
- Anatomy of *my* blood type
- The role of pollutants in a food chain

- Distribution of microorganisms in school areas
- Water quality in a nearby stream
- Lethal effects of radiation
- Book report on a best seller in the area of biology
- Charting my family pedigree (See chart, Figure 4-2)

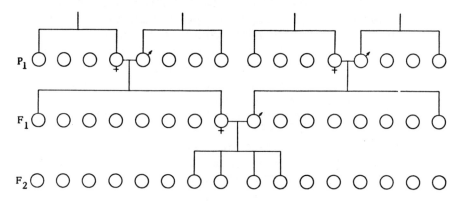

P₁ -- Grandparents and their brothers and sisters
F₁ -- Parents, aunts, and uncles
F₂ -- Yourself, brothers, sisters, and cousins

♂ -- male   ♀ -- female   Ⓣ -- taster   Ⓝ -- non-taster

Add additional lines of relationship where they apply in your case

Fig. 4-2 Sample Pedigree Chart; courtesy of Carolina Biological Supply Company, Burlington, North Carolina

Compiling a Biology Book of Records

Personalized instruction should operate in a warm, human environment in which each individual receives personal recognition and in which each is permitted to engage in meaningful activity, whether alone or as a contributing member of a group. In some situations the personalized interests may be group-oriented, with students contributing on an individual basis. For instance, encouraging students to contribute information about unusual biological incidents or specimens gives an important insight into what kinds of things capture their fancy and may be directed into the compilation of a *Biology Book of Records* (apologies to Guinness), complete with details resulting from researching as well as first-hand observations, investigations, and the amassing of pertinent information concerning rarity or oddity.

Motivation for the personal interest study, of course, should ideally come from the students, but you can help to initiate it by displaying something unusual in a showcase, on a specimen table, in a cage or tank, or on a pegboard or bulletin board. This type of display will give impetus to student reporting (often accompanied by actual specimens, pictures, newsclippings, or other documentation) of similar situations with which they have some familiarity or personal interest. Frequently, they will report some out-of-school experience representative of a personal interest or awareness, such as:

- the largest fish caught in a local fishing area
- the tallest tree in a nearby wooded area or forest
- the largest antlers on a deer bagged during the season
- the longest earthworm found in garden soil
- the tallest tomato plant grown hydroponically
- a frog observed to have jumped an extremely long distance
- a puppy born with extra digits on its paws

Any one of these curious representatives may supply a first-level motivation for an individual to develop a full-scale study in search of answers to his questions concerning the reasons, the causes, and the consequences of the unusual feature identified in his specimen, and to determine if the specimen qualifies for the class project, *Biology Book of Records*.

As the initial interest in the unusual specimen is extended into an investigative study, it may involve making observations and measurements, exploring the size and weight ranges within a species, determining the advantages and disadvantages of a variation within a species, and speculating on the possible effects it will have on future generations. In the course of his personal research and investigation, the student will gain a better perspective of the living world and will become conversant with the application of some of the broad topics and concepts of biology. Those topics concerned with genetic mutations, environmental influence, natural selection, possible evolutionary pathways, the scale effect, and the determination of cell and organismic size limitations are particularly applicable.

Teachers who have encouraged students to become involved in their own learning via personal interests have found that greater learning results when subject matter concepts are developed and applied in a personal context than when they are memorized to fulfill a course requirement or for some other purpose to which the student cannot relate. It is also reasonable to expect that students will reap long-range benefits from the establishment of personalized learning patterns and that they will become life-long learners and better informed individuals.

PROVIDING STUDY OPTIONS FOR STUDENTS

Generally, we find that when we individualize the learning program in its planning phase and make provisions for some student input in the program students tend to learn more biology. Also, we can observe that when the students are given some options and choices in their individual learning experiences, static groups tend not to exist, and, on both an individual and a group basis, more responsible decision-making is in evidence. Thus, from the beginning, each student should find encouragement to elevate his self-image and to assume a positive approach to his investigative studies.

Effective individualized instruction presents options in the form of choices of topics for investigation, direction and depth of study, and types of learning activities for students to engage in. It also provides for adjustments to individual differences in terms of the amount of time required to master a topic study unit, the degree of interest in a topic, the depth of its study to reward the learner appropriately, personal preferences concerning the learning activities that are most suitable and beneficial to the individual, and personal responsibility assumed in a contractual learning situation.

Setting Priorities

The options offered should clearly distinguish the priority levels of various topics and learning experiences so that, in the planning and execution of their learning programs, students can identify that which is most important and in what order topics that are related should be approached. Answers to questions such as, "What would you like most to know about this topic?" and, "What would it be helpful to know before you can investigate this aspect of the study?" will help them to put into proper perspective those studies which are of first priority and those which are extensions of a basic topic. Priority ratings may be designated for topics included in a unit of study. Setting the priorities assists students to identify those topics which are prerequisite to others that seek to discover information and understanding of a more specific nature.

Setting priorities also assists students to maximize the results of their learning efforts and to arrange the sequence of their activities for the most logical order and best understanding. Students have been observed to take a more positive approach to a study when there are clear distinctions of degrees of importance attached to study topics. When time and/or levels of difficulty are also considerations, students tend to apply themselves with greater confidence that they will succeed with optimum results for the time and energy spent if activities are matched with their recognized ability levels. While creativity is encouraged at all priority levels and options are offered to accommodate each individual's interests, aptitudes, personality, and motivational drive, first priority must always be given to the basics, without which there can be no foundation for further learning. A study unit on Energy for Life, for example, can place component topics for investigation in priority categories designated as *basic, suggested,* and *for extended study:*

First priority: *Basic* studies include investigations into Sources of Energy, Energy Levels, Photosynthesis, and Cellular Respiration and Energy Release.

Second priority: *Suggested* studies include investigations into Mitochondria Functions, Aerobic vs. Anaerobic Systems, and Energy Yields.

Third priority: *Extended* studies include investigations into the Effects of Thermal Pollution, Rate of Photosynthesis, Bioluminescence (an energy consideration), Chromatographic Analysis of Leaf Pigments, and the Effects of Temperature and/or Light on Photosynthetic Rates.

Providing Options in Assignments

In any learning program, there may be some students who neglect to prepare assignments or who approach them as burdensome chores. Their reasons are varied but often reflect feelings of inadequacy to understand or deal with an academic requirement, or an attitude that the assignment is uninteresting or lacks challenge. In any case, the value of the assignment, if directed to the "average" student, is seriously diminished. Fortunately, the problem can be alleviated if students are given options which allow them to engage in alternate forms of activity and levels at which they can succeed.

Providing options gives recognition to the fact that there is more than one way to accomplish the learning of a given topic and that students are aware of their individual abilities to learn something better by one method than another. If they are encouraged to suggest the many ways in which a given assignment topic can be developed, their lists will reflect some of their individual talents, strengths, and personal preferences. Since each has an opportunity for input, there will be a provision for every individual to approach the assignment with greater enthusiasm and confidence, to apply himself to better advantage, and, sometimes, to try a new approach—particularly if you convey to him by way of gentle persuasion the message that, "You never know what you can do until you try your hand at something new."

The diversity of approaches and the opportunities for student expression and creativity in a learning experience can be illustrated by two student-prepared lists of assignment options:

Life Cycle of a Frog

Prepare a showcase display of the stages in the life cycle

Compare the metamorphosis of the frog with that of an insect

Collect and preserve a series of specimens showing each stage in the frog life cycle

Research the life cycle in terms of the energy problem and its solution

Read and review one of the following:
Scientific American article offprints:
"Metamorphosis and Differentiation," by V. B. Wigglesworth
"The Chemistry of Amphibian Metamorphosis," by E. Frieden
"How a Tadpole Becomes a Frog," by Wm. Etkin
Oxford/Carolina Biology Reader:
"Metamorphosis," by J. R. Tate

DNA

Construct a model of a DNA molecule

In the lab, extract DNA from *E. coli* or from a fresh calf thymus gland

Prepare a showcase display showing how a DNA molecule replicates itself

Read and review one of the following:
Scientific American article offprints:
"Nucleic Acids," by H. F. C. Crick
"Single Stranded DNA," by Robert Sinsheimer
"Hybrid Nucleic Acids," by S. Spiegelman
Oxford/Carolina Biology Reader:
"Transcription of DNA," by A. A. Travers

Complete the Ward Solo-Learn unit:
"Introduction to DNA"

Research the topic of Recombinant DNA

One teacher I know has reported a marked improvement in student response to homework assignments since she has been offering options in the form of worksheets, drawings, study questions, simulation games, drama for acting out a biological concept, readings in periodicals or selected science fiction, use of special TV, film, filmstrip, and taped programs, and puzzles designed for vocabulary and association building. Her requirement that each student engage in at least six different types of assignment activity during the term still provides a variety of choices but lends encouragement for the exploration of new and different activities. The technique has resulted in her students becoming more versatile in the performance of their learning activities and in approaching the preparation of an assignment with greater confidence.

INDIVIDUALIZING LEARNING TO ACCOMMODATE ABILITY LEVELS

Not all students achieve equally in biology and, while the level of that achievement is influenced by many factors, one of the most important is the ability of individuals to understand, retain, organize, and use information associated with their learning activities. It is when students are presented with tasks at levels at which they cannot succeed that teachers find themselves spending an undue amount of time and energy in remediation, and a difficult situation for both teacher and student follows: one tries to teach and the other tries to learn something that is beyond his capacity for successful performance at that point in time.

Individualizing Learning for the "Slow Learner"

Individualized teaching requires that you have a full repertoire of strategies for providing each student with an optimum learning situation for each task at each level of his development. For example, slow learners, those with short spans of attention, or those who lack sufficient maturity, need to have an instructional program that will take them from their present levels and help them to increase their powers of concentration and depth of understanding, as well as developing their skills for analysis, observation, and creative thought and expression. For these students a full class period devoted to a single activity is generally too long a period of time to command their attention in productive endeavor. There should be a provision for varying the activity to effect a change-of-pace. However, careful planning is needed to ensure a natural flow from one activity to the next as the topic study is developed, and each of the separate activities must contribute to the improvement of the learning patterns and processes of the individual students.

A lesson which investigates the flight reactions or the feeding habits of a housefly can be developed in a single 40- to 45-minute class period, divided into five short sub-periods that are devoted to five different activities of the learning experience:

1. MOTIVATION (5 minutes) — Upon entering the classroom/laboratory, the students: arrange to work individually or in established groups of three to four students; refer to guide sheets stating the investigation for study; and examine the materials and specimens to be used in the investigation.

2. INVESTIGATION (15 minutes) — Using guide sheets, students follow simply stated instructions. ("Glue cold-immobilized housefly to wooden stick applicator. Observe responses as revived fly is placed in different positions while held aloft at tip of stick and/or reactions of mouthparts when in contact with a drop of sugar-water on a hard, non-absorbent surface.")

   During this time, the teacher will circulate and visit each work station to assure students that the learning experience has importance. Students will proceed with guide sheets and make records of observations. In a group situation, teamwork is in order, and one student may act as a recorder of information agreed upon by the group.

3. COMMUNICATION (10-15 minutes) — Students assemble for a discussion in which all individuals and/or groups share information, raise questions, suggest other situations where this information can be applied, and relate personal experiences or reading references related to the topic.

4. REINFORCEMENT OF LEARNING (5 minutes) — Each student writes a short summary in which he states his understanding of the information gathered via the investigative study.

   OR

   Each writes answers to questions in the form of a short quiz based on the investigative study.

5. FOLLOW-UP (5 minutes) — A short summary of the investigative study is followed by a look ahead to another phase of

the overall topic. Specific aspects of the continuing study are accepted by volunteers or assigned to individual students, often selected because they have raised related questions during the discussion period.

The designated time periods are, of course, flexible, but the format is excellent for slow learners and immature students. It helps them to build confidence that they can succeed because it focuses in on precisely what is important in a series of meaningful activities, none of which is lengthy enough to cause them to lose interest or to resort to day dreaming. It ensures their complete involvement in a biological investigation and increases their enthusiasm for learning, while omitting what, for them, would be meaningless detail in the form of names of insect mouthparts, scientific names, and classification terminology.

Individualizing for the College-Bound Student

The college-bound student should also be given an individualized program that is strong on exposure to many and varied experiences. His program should involve: the gathering of important biological information from authoritative sources, current references, and research journals; experimenting and researching in the laboratory as well as in the library and media center; writing reports; making observations; organizing and analyzing data; and arriving at valid conclusions based on the evidence. These experiences should take him beyond the classroom, library, and laboratory, and expose him to the realm of field and research laboratories where he can visit biologists at work. His contacts with many resource persons—teachers, biologists, and visiting lecturers from related industries and colleges and universities—should prevail.

Studies indicate that students who have had a thorough grounding in General Biology, which emphasizes the major theories of the Cell, the Gene, and Evolution, are better prepared for college than are those who may have studied only specialized biology while in high school; and that interest in biology heightens after the student has become involved and has mastered the fundamentals. Consequently, college-bound students need to have the broad familiarity with fundamentals of biology that will give them the perspective they must have before they can attempt to understand a specific organism, a relationship, or a specialized biological topic, whether pursued while in high school or as a follow-up in college.

Basic training and preparation for the more specialized biological fields is enhanced by activities which provide experiences in quantitative observation or biological measurement in an interesting and challenging context. It is through studies involving observation, measurement, tabulation, and interpretation of data, and application of findings about significant factors relating to biological systems that patterns are established for specialty studies. Some of these observation-measurement studies include effects of:

- temperature on CO_2 production by a yeast culture
- pH and inorganic additives on catalase activity of liver cells
- pH on formation of coacervates
- UV light on growth of microorganisms
- light intensity on chloroplast activity in Elodea cells
- cell size on rate of diffusion through cell membranes
- fatigue on contraction of skeletal muscle
- height of individual on vital lung capacity

Individualizing for Advanced and Gifted Students

Advanced and superior students can be given greater responsibility for their own learning programs because they are generally highly motivated and capable of engaging in independent study. To help them realize this potential, highly challenging work should be provided, with as much freedom for its development as they can properly manage. There is, however, one word of caution: Take care that you do not permit a program for a highly capable student to become one in which he finishes quickly and is left with nothing productive to occupy his time, or in which he merely engages in more work instead of in work that is more challenging.

Individualizing for Developing Awareness of Career Opportunities in Biology

Biological studies can take on a new dimension and become more meaningful for students if the course is infused with career relevance which is directed toward individual interests and abilities. As an alternative to a separate unit devoted to Careers in Biology, the same components that might otherwise be confined to a 1–2 week period can be integrated with other learning activities throughout the term. Each student is thus afforded many opportunities to respond—on an individual basis, at whatever point in time, and to whatever stimulus motivates him—to biology as a possible career. Consequently, each can be assisted in reaching a decision concerning whether or not to pursue an interest in biology as his life's work.

Careers in Biology might well be a unifying theme, threaded throughout the course. Its purpose might reasonably be achieved by employing the following simple guidelines:

- Plan activities and learning experiences that will enable students to develop an awareness of the total spectrum of careers in biology.
- Be alert to topics and activities that attract individual students and suggest that they engage in extensions of these interests by way of open-ended investigations.
- Encourage students to attend career conferences relating to their identified interests and/or abilities in biological studies.
- Invite former students who are now training for or engaged in biology-related careers to relate pertinent information to current students who are considering similar careers.
- Plan field trips to research laboratories and other facilities to allow students to view professional, semi-professional, and technical personnel engaging in many different phases of a particular field of biology-oriented work.
- Post brochures and other literature from colleges and training schools that offer biology-oriented programs, and encourage interested students to make use of the request cards for obtaining further information and details.
- Encourage students to gather information on personal qualifications, training and educational requirements, starting salary, opportunities for advancement, trends and projections for the future, etc. about a career being considered, and to make a personal assessment to determine the suitability of the career for the individual.
- Where possible, put individual students in contact with a practicing biologist for the experience of working on a project or for absorbing some of the atmosphere of a biologist at work.
- Encourage students to train for the future and for a *field* rather than a narrow specialty in biology; help them to gain a perspective of the importance of breadth, scope, and flexibility for professional success, as well as the importance of readiness to make adaptations as priorities change.

INDEPENDENT STUDY AND RESEARCH PROJECTS

Research at the high school level is usually limited to advanced and gifted students. Its concern is with the search for information not already known or understood and does not include the performance of a standard experiment that merely places the student in the position of discovering for himself what will happen. While it implies a laboratory-type study conducted by an individual in a completely independent manner, there are certain elements of a student research program which point up responsibilities for both student and teacher. To ensure the success of our program, my students and I have set up guidelines that allow us to fulfill them cooperatively:

1. *Motivation* should ideally come from the student, but a personalized interest may also be detected by the teacher, who then refers the student to further investigation of the area in order to identify a topic for research.

2. *Attitude* is that of open-mindedness and acceptance of responsibility for proper application of the scientific method while seeking truth and understanding.

3. *Readiness* is determined by the exhibited capability of the student to apply scientific procedures and only after proper background about the topic has been gathered from extensive library research and other authoritative sources.

4. *Time* is provided for the research project by the waiving of some other course requirement.

5. *Materials* and *resources* are located and made available for student use. Where possible, those not a part of the school's facilities are either obtained on a loan basis or made accessible for use at a cooperating local hospital or research laboratory.

6. *Interest* and *support* are shown in the student's research project. Rapport is established during conferences and informal visits throughout the planning, working, and interpretative stages of the project. Where possible, the student is put in contact with other teachers and practicing biologists who, because of their expertise, are in a position to function as additional resource persons.

CREDIT FOR OUTSIDE WORK

We all are aware that many students engage in biology-related activities outside of school and that the learning experiences concerning the life sciences are not confined to the classroom/laboratory activities.

Many 4-H and Audubon activities, part-time and summer jobs in pet stores, greenhouses, and fish hatcheries, and family and individual hobbies such as breeding dogs, cats, and parakeets, maintaining a fresh or salt water aquarium, or growing plants of unusual varieties can offer rich experiences which are often accompanied by expert guidance. Whether a part of a directed study, such as tagging animals to investigate their migratory patterns, or unplanned situations encountered during the normal course of raising plants and animals, these experiences can become an important part of the individual's learning program and receive recognition in the form of credit.

Each student's case is, of course, highly individualistic, and conferences with the student must be held in order to assess the nature of the activity, the amount of time devoted to it, and the kind and amount of knowledge and understanding resulting from it. Planning with the student, it may be possible to extend the study to include some aspects not touched upon by this outside experience, or, if the study appears to be self-contained, credit may be given by the waiving of a course requirement that appears to be of equivalent value. Some teachers have accepted the outside work as an alternative for an individual project study, normally a part of the course, and others have given bonus points to be added to those earned by the student as he meets point qualifications for a given grade. In all cases, there is a recognition that the outside work contributes to the student's learning of biological information and that the interest engendered by it will probably remain active, particularly if its importance is recognized and rewarded.

ROLE OF THE OPEN CLASSROOM

The atmosphere in which students engage in their individualized learning programs is an important factor in the determination of how effective that learning will be. The involvement experiences, problem solving, and learning by discovery that are necessary for the development of abstract biological concepts cannot be accommodated in a traditional recitation-oriented classroom. Nor can the variety of learning experiences for providing individualized learning programs all be accomplished within the confines of the allotted time of a standard class period. Both the classroom and the laboratory must be structured and available according to an open plan. Priority must be given to their flexibility in use so that the space, divided into interest, resource, and activity centers, is adaptable to easy rearrangement for discussion sessions, and the facilities are readily available at times which are convenient for students who wish to avail themselves of the opportunities to engage in learning experiences. While the emphasis is on *independent* learning, the presence of the teacher, who confers with individuals on a one-to-one basis and acts as an interested resource person, is essential to its effectiveness.

Some teachers keep a log book in which students record their time and activity devoted to an individual or self-initiated learning experience in the open classroom. Another has found satisfaction in the use of personal student journals in which:

- no distinctions are made between classwork, lab work, and homework,
- students schedule themselves for activities in the open atmosphere as they progress in the development of their individualized programs,
- each student enters in his journal a summary or work sheet for each activity completed, and
- students may refer to their journals for writing a final summary report when the study unit has been completed.

This teacher's satisfaction reflects basic student satisfaction with the way an open classroom situation encourages each to assume the responsibility for his own study in an atmosphere that allows him freedom to engage in his learning activities at his own convenience, but with the security of an interested resource person close at hand.

The experience of a teaching colleague who has attempted individualized teaching in both a traditional and an open classroom confirms what we would expect: "that through freedom to utilize the learning style best suited to his individual needs, each student was more effectively able to complete his learning program and, through the availability of the open room, he seemed to attach a greater importance to his learning." The willingness and eagerness to use the open classroom facilities during non-scheduled hours seems to attest to the importance each student placed on his own learning. Through it we can hope that the student is moving toward greater maturity and respect for learning as a lifetime activity.

USING CONTRACTS

Making a contract between yourself and individual students encourages to the utmost the development of student responsibility. As another alternative to a traditional approach, it involves autotutorial instructional methods and is particularly well-suited for the highly-motivated, the scientifically-oriented, and honor students who are high achievers.

In individually conducted conferences—during which agreements are reached concerning student objectives for a given unit of work, development of a program for reaching those objectives, dates for completion of the work, and establishment of a personalized performance evaluation standard—each student "contracts" with you to produce a certain kind and amount of work for a given grade. Generally, requirements for each grade are correlated with achievement of the objectives and with the depth of study involved. Variations of a basic grading system may be used:

"C" for completion of a basic program study and mastery of subject matter and its application as determined by test scores of 85% or better,

"B" for completion of an expanded program, test scores of at least 90%, and an in-depth investigative study of a topic to be reported in a research paper and/or in oral and demonstration form to a designated group,

"A" for completion of an extended program, test scores of 95% or better, and an extensive investigative study in the form of an original research project of quality suitable for entrance into competition in a local, state, or national talent contest.

The contract approach allows the student great freedom while demanding that he assume great responsibility; it permits for flexibility in the selection and development of topics for study, allows a student to choose the grade he desires and is willing to work for, recognizes outside work, accommodates personal interests, encourages resourcefulness, and always keeps him aware of his part in fulfilling the terms of the contract agreement.

Many who have used the contract plan are enthusiastic about the accompanying increase in student initiative and input but also agree that, in the interest of helping students to develop responsibility to the fullest extent, the program must be carefully monitored. Specifically, it is helpful for you to do the following:

- Make sure that the objectives and terms of the contract are clearly stated and understood.
- Assume a flexible posture but agree only to objectives that to you are personally and professionally acceptable.
- Confer frequently with individual students concerning their progress.

- Remain available to help students when they have questions or are in doubt.
- Function in the role of an interested resource person.
- Encourage students to upgrade their contracts, if, during the course of events, it appears feasible.

Students learn as individuals and cannot be fitted into a mold by which a predetermined kind and amount of biological knowledge is learned in a predetermined time period, by a predetermined method, for the same predetermined purpose, and to some predetermined degree of acceptability. To design learning programs that are geared to individuals, we need to incorporate well-ordered flexibility into a learning environment that is in harmony with their personal needs and interests as well as their individual ability levels. Options for students may very well provide the solution—they offer the individual a flexible working program while actively involving him in determining the direction that his learning will take and the degree of success and satisfaction that he will ultimately achieve.

5 Selecting and Using Effective Demonstrations

Whether we teach in a standard or modified traditional classroom/laboratory setting or in an open learning atmosphere, we all find it desirable, on occasion, to have all students focus their attention on a central theme. In the motivation for study of a new topic area, the illustration of a basic biological principle or natural phenomenon related to a study under way, the introduction of a technique to be used by students in an investigative study, or the explanation of a technique employed as part of a research procedure for gathering information found in published reports, a demonstration can be far more effective than the mere telling. If it is true that a picture is worth 1000 words, then surely a good demonstration must be worth 10,000.

Biology demonstrations can be conducted with or without student assistance. In some cases, a highly-motivated, well-prepared student who has thoroughly investigated a topic and perfected the technique for its demonstration to others can give a creditable presentation. But, however and by whomever conducted, it is important that each demonstration be true to its purpose—each should help students to visualize a concept that might otherwise be difficult to comprehend; each should illustrate without being directly involved with inquiry and investigative activities; and, since its design is such that few students have an opportunity for active involvement in its performance, each should capture and hold student attention while engaging them in mental activity that is designed to ensure their comprehension, belief, and/or understanding of the topic. Well-planned, well-conducted demonstrations offer capsule learning experiences; they can add an important dimension to the teaching-learning situation in your classroom.

Because of their extensive exposure to the influences of the world of entertainment, advertising, and T.V., students have gained a certain sophistication concerning programs and presentations and they are very discerning in matters of content, technique, timing, delivery, development, and relevance. In order to compete with these strong influences, a teacher demonstration must meet rigid standards. It must be well-planned, well-executed, well-polished, and well-directed. It must also hold together as it progresses from introduction to conclusion, and with a precision that flows naturally while it commands attention and directs thought toward a worthwhile, meaningful, and relevant goal. Further, yours must go beyond the characteristics of a professional "performance"—yours must also stimulate student mental processes and encourage application of a single concept to other aspects of the central topic and to the total learning program.

ENSURING PROFESSIONALISM IN TEACHER DEMONSTRATIONS

Many factors involved in the planning, preparation, execution, and follow-up of a biology demonstration contribute to its success and effectiveness, and, in judicious combination, can distinguish yours with a touch of the professional. To achieve this stature in a lesson, nothing can be left to chance, nor can anything be allowed to detract from the demonstration's main thrust.

Guidelines for Ensuring Professionalism in a Teacher Demonstration

- Select topics to be demonstrated with care and limit any given demonstration to one principle or one phenomenon. Don't confuse students by trying to develop more than one concept at a time.
- Research all aspects of a demonstration topic thoroughly so as to enable you to take off in any direction related to the topic.
- Prepare broadly as well as specifically on the theory and applications of the demonstration principle and/or phenomenon. Be prepared with proper background to relate it to other areas and to project a resourceful approach which suggests open-ended studies for individual research and investigation and/or in-depth study of a related topic of personal interest.
- Practice the demonstration to ascertain its time requirement, to assure continuity and smooth delivery of an interesting and well-rehearsed commentary, and to develop self-confidence and competence to be able to handle an emergency (should one arise) in a resourceful manner. There should be no awkward pauses, no indecision, no un-

planned surprises, no rushing to complete an unfinished demonstration as the period draws to a close, and no need to offer excuses, such as "This is the way it was supposed to work."

- Assume a personal responsibility for making or supervising all advance preparations:

    —Check specimens and/or cultures, if needed, to be sure of proper age, condition, and suitability for use in the demonstration

    —Check equipment to ascertain its good working condition and your complete familiarity with all systems of operation

    —Retain appropriate samples of preliminary, intermediate, or resultant stages from a practice run for use in a demonstration which, due to time limitations, must rely on a "time-lapse" technique

- Keep each demonstration short and simple in design. Simplicity is a key point. Not only can basic principles and natural phenomena be best demonstrated when left uncomplicated by elaborate and intricate design and operation, but a simple presentation will usually work better and more reliably; the use of simple equipment allows for full concentration to be focused on the demonstration itself rather than on an ingenious instrument and its manipulation, and explanations that are direct and interesting are more effective than those which employ terminology that is overly technical and refers to things with which students have little or no familiarity.

- Prepare an interesting introduction or discussion to motivate interest in the demonstration which relates to a topic under consideration, establishes students' previous knowledge of the topic, and emphasizes the importance of the principle involved.

- Prepare and distribute handout sheets listing some of the significant points to observe and providing space for note-taking and for summary and concluding remarks.

- Prepare a follow-up quiz, discussion, or assignment to reinforce the learning experience with understanding of the concept demonstrated.

- Allow and specifically arrange for some student input—some may participate by assisting in the mechanics of the demonstration and all should be involved in mental activity via thought participation, problem solution, and speculating on "what if?" situations suggested. Welcome and encourage extensions of the demonstrated principles and applications to new situations suggested by students for independent or follow-up study.

- Select a proper vehicle for conveying the demonstration to students. Clear viewing by all students must be assured, whether the presentation is in person to a small group or to a larger one via closed circuit T.V., live or on tape.

CLOSED CIRCUIT T.V. FOR DEMONSTRATIONS

Demonstrations which give students information and techniques necessary for employment in their investigative laboratory work, which involve many time intervals between successive stages of a single phenomenon, and/or which are enhanced by the viewing of small objects or specimens by close-up or magnification techniques, are most effectively presented via the use of a closed circuit T.V. system. Particularly in a large class, each student is able to see a demonstration more clearly on a viewing screen than would be possible had it been performed at the front of a large classroom or with all students crowding around a demonstration table or work bench. If it is possible to resolve scheduling matters, T.V. may also allow for a single presentation to serve the needs of several classes meeting simultaneously and, if the program can be taped as well, it will provide material for students to use for replay, either for review or for first viewing of demonstrations missed due to absence. With its many advantages of convenience and effectiveness, this mode of presenting demonstrations is both popular and well-received by students.

If there are several biology teachers in your school, you may want to consider a cooperative endeavor for presenting information via closed circuit T.V. This works well in situations where more than one teacher plans to use similar demonstration materials at the same time or in a similar manner. By a pooling of talent and expertise, each is relieved of the responsibility for *all* demonstrations and there is an opportunity for each to contribute in areas of special interest and strength. By devoting a greater amount of concentration, time, and effort to the preparation of a smaller number of select demonstrations, each can turn out masterful presentations that are truly superior. Where employed, this cooperative enterprise has been reported as very successful; students attach greater significance to the more carefully conducted learning experiences, and teacher time and effort are utilized more advantageously and professionally in the process.

The mechanics of televising and/or taping will, of course, be determined by the nature of the equipment available. If possible, two cameras, both controlled by the teacher, should be used: camera 1, focused on the instructor and her desk or table for overall viewing and long shots, and camera 2, for close-ups and detailed viewing of pertinent living specimens, microscope views, charts, diagrams, and fine points of examination of materials employed.

DEMONSTRATIONS THAT WORK

Many demonstrations can be completed in a short space of time and serve as a dramatic focal point of an overall lesson or study topic. There are some techniques that can be completed in 10 minutes or less, while others, in person or on closed circuit T.V., require more time to develop. Whether limited to a single class session or viewed on an on-going basis over an extended period of time, your demonstration should be used to transmit a minimum amount of information to your students, while guiding them toward a maximum of mental participation in the activity. Demonstration techniques can be varied, but the approach should always be direct, simple, and with lean explanations that will stimulate student thought processes and help each to attach meaning and understanding to the concept, principle, or phenomenon demonstrated. Some of the numerous demonstrations whose designs meet these criteria are described below.

A demonstration of the source of oxygen breathed by fish in water: Boil 400 ml. of water in each of two beakers and allow to cool to room temperature. Aerate the water in one (A) but not the other (B). Select two healthy goldfish of equal size, but with some identifiable marking to distinguish them, and place one in each beaker. Allow students to observe both fish, and call attention to the gill activity. After students have noticed the difficulties being experienced by the fish in non-aerated water (B), quickly rescue the "fish-in-distress" and transfer it to the beaker containing aerated water (A), in which the fish originally placed is still thriving. Call attention to reactions of the distressed fish as it revives, and discuss the difference in conditions in beakers (A) and (B) and the demonstrated need and source of oxygen. It should be clear to all students that the oxygen which fish breathe is that which is dissolved in the water and not the oxygen atoms that are chemically bonded to hydrogen in the water molecules. As a follow-up, students may investigate the rate of operculum movements and its relation to the physiological state of the fish; or they may pursue some broader aspect of the effects of oxygen-depletion in water, as suggested by the demonstration.

A demonstration of transpiration by plant leaves: Select a 16- to 18-inch length of leafy branch with a stout stem and make a sharp diagonal cut at its lower end. Attach one end of a 14- to 16-inch piece of rubber tubing tightly over the cut end of the branch, and the other end, with a tight connection also, to a graduated 10 ml. pipette. Immerse the pipette and tubing in a shallow pan of water so that both become completely filled with water as all air is displaced by a continuous column of water in contact with the cut end of the branch. Orient the branch and the pipette in a fairly vertical position and, with masking tape, attach them to a window or similar surface in a lighted area, allowing the rubber tubing to curve gently between them in a near horizontal position. Students should observe and determine the cause of the movement of water as it adjusts to different levels in the pipette. Follow-up investigations of a plant's ability to regulate this process, what happens when a plant is defoliated, and comparative studies using branches from several different species may be pursued.

A demonstration of O_2 production during photosynthesis: Prepare two test tubes, each half filled with water, and place a sprig of healthy Elodea in one (A) but not the other (B). Fit each tube with a cork through which a large dissecting pin has been inserted and on which a ¼-inch piece of white phosphorus has been impaled so that it is suspended above the water level in the tube. Place both tubes in a position where students can observe the white fumes emanating from the phosphorus as it oxidizes where exposed to oxygen in the air spaces above the water in the tubes. Explain the oxidation of phosphorus reaction and note the reasons why the emanation of white fumes eventually comes to a stop. When this occurs place both tubes in front of a strong bright light source or in sunlight and have students observe, within 8 to 10 minutes, a resumption of the production of white fumes in the tube containing the Elodea. They should associate the production of oxygen causing the phosphorus to oxidize with the green plant and with photosynthesis. A follow-up which includes the biochemistry of photosynthesis, the importance of green plants to animal forms in aquaria, ponds, lakes, and other aquatic or marine habitats, and the importance of photosynthesis to a balance of life forms on earth will provide individual topic studies for both laboratory and library research.

A demonstration of the activity of an oxidative enzyme found in fresh produce: In a Waring Blendor macerate 10 g. of fresh mushrooms in 80 ml. of cold water and filter the resulting material through cheesecloth. Transfer 9 ml. of the filtrate to each of two large test tubes (A and B), and observe its color. To one of the tubes (A) add 9 ml. of a solution of catechol (solubilized in alcohol and adjusted to a concentration of 0.1% by the addition of distilled water) and mix well by agitation. Shake both tubes vigorously for 5 minutes and again make a color examination and comparison of the two tubes. Students should associate the color change with that occurring in cut apples, potatoes, and other fresh produce. The widespread distribution of the polyphenolases (such as catecholase) should be established and observations of this phenomenon primarily in exposure of fresh, but not cooked, fruits and vegetables to air reported. The demonstration can be used to introduce or to review and reinforce a classic student laboratory investigation of catalase in fresh and boiled liver.

A demonstration of the bactericidal action of lysozyme: Add 1 ml. of a water dilution of fresh egg white to a culture tube containing 10 ml. of a broth suspension of *Sarcina lutea*. Note the clearing up of the turbidity of the bacterial culture, while a second, untreated suspension of the same species retains its characteristic turbid appearance. This should be recognized by students as a lysozyme action of protein digestion and related to lysosomes of animal cells in which protein-digesting enzymes are stored and often used against invading bacteria.

TIME-LAPSE DEMONSTRATIONS

Sometimes it is desirable to demonstrate a phenomenon in which a period of time must elapse between the preparation and the outcome—both of which are important for a full understanding of the principle involved. In these cases, a pre-run of the demonstration can serve a dual function: (1) it can provide the necessary practice for total familiarity with the demonstration techniques; and (2) it can supply appropriate stages of the procedure, allowing them to be introduced in proper sequence via a time-lapse technique, when the demonstration is presented in a single uninterrupted session. A few examples of demonstrations of this type include the following.

A demonstration of the sensitivity of bacteria to U.V. light: Heavily inoculate a sterile nutrient agar plate by swabbing its entire surface with a sterile cotton applicator dipped in a 24-hour broth culture of *Serratia marcescens*. Place a 2½-inch white paper letter on the surface of the inoculated agar, making certain that contact is made with the agar and, while wearing protective goggles, place the entire assembly (with cover removed) under a U.V. lamp for 15 minutes. After exposure, remove and discard the paper letter, replace the cover on the petri dish, and incubate the covered plate for 24 hours at 37°C. Examine a similarly prepared plate from the previous day's pre-run, and allow students to observe the characteristic red pigmented growth of the organism in the shape of the letter which had acted to shield it from the U.V. rays, while those in the unprotected areas were killed. The newly prepared plate can be allowed to incubate and will serve, 24 hours hence, as a confirmation of the demonstrated bactericidal action. An investigative study into bacteriostatic and bactericidal agents and their practical applications can be used as a follow-up study.

A demonstration of phototropic response in euglena: Fill two vials with a rich culture of euglena and place opaque screw caps in position. Cover one vial (A) completely with thick black paper and cut a narrow slit lengthwise in the paper to admit light. Do not cover the other vial (B). Place both vials in sunlight or near a bright artificial light source so that the slit in (A) faces the light. Examine vials that were similarly placed several hours earlier and remove the paper cover from (A) to reveal the green streak of euglena clustered along the line where the slit in the paper had been positioned. Students should understand the relationship between the slit in the paper and the source of light to which the euglena responded in a positive manner. The relationship between the favorable phototropic response of this species and its survival can be explored further, and some possible evolutionary pathways can be traced.

A demonstration of sensitivity of a bacterial species to antibiotics: Cool a flask of melted, sterile nutrient agar to 45°C and heavily inoculate it with Bacillus subtilis from an 18-hour slant culture. Distribute the organisms evenly by gentle agitation, and pour the inoculated agar into a sterile petri dish and allow to solidify. When firm, spot the surface in well-separated positions, with a series of antibiotic disks, representative of a spectrum of antibiotics. After making sure that contact has been made between the disks and the agar surface, cover the petri dish and incubate, in an inverted position, for 18 hours at 37°C. Examine a similarly prepared plate from a pre-run of the demonstration and allow students to observe the halo-like clear zones surrounding some of the disks, standing out in sharp contrast to the lawn of bacterial growth on all other areas of the surface. Measurements of the zones of inhibition can be taken as an indication of the degree of sensitivity of *Bacillus subtilis* to each of the antibiotics and, conversely, the relative effectiveness of each antibiotic against the organismic growth of this species. Students should gain a clear concept of antibiosis and of the effectiveness of antibiotics against a specific species of microorganism. In a follow-up study they might explore other agents, other concentrations, and other organisms of sensitivity or resistance. Cases of natural an-

SELECTING AND USING EFFECTIVE DEMONSTRATIONS   113

tibiosis might also be investigated by interested students, who can then report their findings to the group. The freshly prepared plate with antibiotic disks should be incubated for a later examination and confirmation of the demonstrated results.

A demonstration of the protein digesting action of bromelin: Place two or three 4-inch lengths of capillary tubing in freshly made raspberry Jello, mixed according to directions on the package. After the liquid has solidified in the tubing, remove the tubes from the Jello bowl and clean off excess material adhering to the outside surfaces. Using a triangular file, cut the Jello-filled tubing into 1-inch sections and place two such sections in a test tube containing enough pineapple juice, maintained at room temperature, to cover completely the small capillary sections. Similarly prepare a second test tube, but with pineapple juice that has been boiled. Examine the results of a pre-run of the demonstration and allow students to observe that the Jello has dissolved in the unheated pineapple juice. An examination of Jello-filled capillaries placed in the heated juice can serve as a comparison study, and the freshly prepared tubes can be left for a confirmation of the demonstrated results. Students should identify bromelin in the pineapple juice as a protease, the activity of which is destroyed by heat.

DEMONSTRATIONS FOR PROJECTING OR TAPING

In some situations in which classes are large and specimens are small or in need of magnification, better viewing is made possible if projection techniques, live with an overhead projector or taped for projection on a viewing screen, are employed. It ensures that all students will see the same view of a microscope study and that no student will be handicapped by the viewing of something which is out of focus or otherwise unsuitable for the purpose intended. Demonstrations of laboratory equipment and techniques particularly can be enhanced by taping because of their availability for replay and for reviewing by students engaged in independent laboratory-oriented studies. Seven demonstrations which lend themselves well to these criteria are described below.

1. *Demonstration of the mercury amoeba:* Place one drop of clean mercury in a syracuse watch glass and cover with dilute nitric acid. While this is being projected for viewing, add a few crystals of potassium dichromate to the acid near the mercury. Students should observe the

amoeba-like movement and from this develop a concept that a definition of life must go beyond the ability to move. The demonstration can lead to a very lively discussion and consideration of "What is life?"

2. Demonstration of the feeding response in hydra: Transfer unfed hydra to a clean syracuse watch glass and observe under magnification of the dissecting microscope. Add brine shrimp that have been rinsed free of salt and observe the hydra's capture and ingestion of the prey. Using fresh hydra, add 1 drop of glutathione solution to the culture and observe the feeding response again. With fresh specimens once more, offer a piece of a freshly formed dry blood clot on the tip of a pair of forceps to a hungry hydra. Students should be able to analyze and interpret the feeding response demonstrated.

3. Demonstration of circulation in a fish tail: Wrap the body of a goldfish in wet absorbent cotton, leaving only the tail exposed. Place the fish in a petri dish containing a small amount of water and hold the tail flat and expanded by covering it with a microscope slide. Examine under low and high power of the compound microscope, and observe the circulation of blood in the arterioles, capillaries, and venules. If necessary, use applications of ether, urethane, or chloretone to quiet the fish, but keep in mind that the use of an anesthetic will cause the circulation to become sluggish.

4. Demonstration of ciliary streaming: Slit a fresh chicken trachea from its lower end upward, being careful not to damage the delicate lining except along the line of the incision. Pin the trachea to a block of paraffin, spreading the incision so that the inside surface can be viewed as a fairly flat surface, and place the paraffin block in a tray of water with the upper end of the trachea somewhat elevated. Place a small drop of India ink on the mucous membrane at the lower end of the trachea and cover the entire assembly with a beaker inverted in the tray. Observe the movement of the ink and, using a mm. ruler and a stop watch, determine its approximate rate of travel upward, against the force of gravity. Students should associate this phenomenon with the cleansing action of cilia moving against gravity in the trachea where dust, if inhaled, will be swept up and out. Agents which may be detrimental to this action should be researched.

5. Demonstration of technique for observing internal functions of a crayfish: Place an active crayfish in a battery jar or beaker of suitable size and cover completely with club soda to anesthetize. When the specimen no longer responds to gentle prodding, carefully remove its carapace to reveal the functioning dorsal heart and lateral gills. One such treated specimen will last for hours and, since its preparation simulates the natural process of molting, it has no upsetting effect on students.

6. *Demonstration of contraction of skeletal muscle fibers:* Tease apart individual fibers of a 3- to 4-mm. section of frozen glycerinated psoas muscle on a glass microscope slide containing a drop of glycerine. Mount one fiber on a fresh microscope slide and examine under the dissecting microscope to determine its length and appearance. Add several drops of 0.27% ATP in 0.05 M KCl + 0.001M $MgCl_2$ in distilled water and, after 30 to 60 seconds, observe the contracted fiber under magnification of a compound microscope. The observed differences in structure should be associated with the shortened length, and the physiology of muscle contraction should be explored in greater depth.

7. *Demonstration of osmosis:* Immerse an egg in white vinegar with 5% acidity and, when the shell has been completely dissolved, carefully transfer the deshelled egg to a beaker of water. After 3-4 hours, remove the egg from water and weigh it. Transfer the turgid egg to Karo syrup for 24 hours and weigh it again, noting its flaccidity. Once more immerse the egg in water for a 24-hour period and note the restoration of turgidity. Students should have a clear picture of the role of a membrane in controlling a cell's (the egg's) turgidity and be able to relate the phenomenon to cells of both unicellular and multicellular organisms used in their laboratory and independent investigations.

DEMONSTRATIONS FOR DISPLAY

Many topics can be effectively demonstrated in the form of a display set up on a desk or table, or mounted on a bulletin or peg-board where students can refer to the "demonstration" of some aspect of a topic study under consideration. Charts, illustrations, models, and ditto sheets to accompany the demonstration can be prepared and made available for student reference. Display demonstrations are best when kept simple, as is illustrated by the following examples.

Demonstration of conduction in plant tissues: (1) Immerse the split stem of a white carnation in colored ink—one end in blue, the other in red. Display the specimen and allow students to observe the resulting conduction of these two colored substances from the beakers to the flower petals. (2) Immerse a celery stalk in blue ink. Cut a cross-section of the stalk and place it for examination under a dissecting microscope. Students will be able to see visible evidence of conduction of the liquid in the conducting tubes.

Demonstration of stages in bud development: Collect buds from a tree or flowering bush in spring, one bud per day for a period of time spanning first appearance to full bloom. Preserve each bud in a separate jar of formaldehyde and display, in sequence, for students to note the progression of stages in the opening of a flower bud.

Demonstration of stages in frog embryology: Collect and preserve in individual jars of formaldehyde samples representing cleavage stages, blastula, gastrula, tadpole, metamorphosing individual, and adult frog specimens and place in numbered sequence for students to view under a dissecting microscope.

STUDENT INVOLVEMENT IN DEMONSTRATIONS

There are some demonstrations which can involve students as active participants, singly, in small groups, or as an entire class. In a hands-on situation, students are frequently more responsive because of the greater degree of personal involvement in the demonstrated phenomenon. After having incorporated more student involvement in biology demonstrations, my colleagues and I found that concepts were being mastered in a shorter period of time and the knowledge gained from the learning experiences was of a more permanent nature, with greatly improved retention of both factual information and its application to new situations. Good examples of demonstrations using the three types of student involvement include the following.

Demonstration Involving All Students

1. *Location of a blind spot:* Prepare a diagram (as shown in Figure 5-1) in which a ¾-inch blackened cross is placed 1-inch to the left of the midpoint of a white index card, and a ½-inch circle is similarly placed to the right so that both figures are at the same horizontal level but separated by a distance of about 2 inches of white space. With the left hand cover the left eye, and with the right hand hold the card at arm's length in front of you as you focus with your right eye on the black cross. You will be able to see the circle as well. Slowly move the card closer to your eye until you reach a point at which the circle disappears from view. This is your blind spot. Moving the card still closer (or uncovering your left eye) brings the circle back into view. This demonstration is very effective as a prelude or as a follow-up study of a laboratory dissection of a cow or pig eye for structural detail.

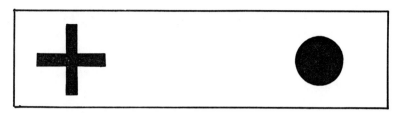

Fig. 5-1 Diagram, drawn on a white card, for demonstrating the "Blind Spot"

2. *Eye dominance:* Using both hands, hold a 2-inch diameter rolled-paper telescope about 6 inches in front of your eyes and look through it at some object across the room. Holding your hands steady, close first one eye and then the other. The eye which sees the object through the tube demonstrates which of your eyes is the dominant one.

3. *"Hole-in-the-hand":* Hold a 2-inch diameter rolled-paper telescope in front of your right eye, using your right hand to hold the 'scope and your right eye to focus on the wall clock. Keeping both eyes open, place the outside edge of your left hand, with palm outstretched and facing you, alongside the edge of the telescope, just a few inches in front of your open left eye. You will now be able to see the clock through a circular hole in your hand.

4. *After-image:* Look at a bright light for several seconds and then glance away. You will see an after-image, sometimes very colorful. Gaze intently at an outline of the American flag on which one black star on a yellow background and alternating green and black stripes have been drawn. After 60 seconds look at a white background and see a red, white, and blue flag as an after-image!

5. *Articulation of a chicken foot:* Obtain a chicken foot from a butcher shop and cut the upper end so that each tendon is free and can be grasped by its cut end. Tug on each tendon, one at a time, and observe which toe is moved by each.

6. *Cytoplasmic streaming:* Examine agar plates on which *Physarum polycephalum* sclerotia had been placed 24 hours earlier. Note the spread of pigmented substance beyond its original placement position and examine under magnification of the compound microscope for streaming of the granular cytoplasm.

7. *Demonstration of absorption of CO_2 gas produced by fermentation:* Place 1 ml. of a 1% KOH solution in the fluid contained within the bowl of a Smith fermentation tube that shows a column of gas collected

during the fermentation of sugar by a yeast culture. Hold your thumb over the mouth of the Smith tube and watch the liquid in the bowl rise back into the gas-filled column as the CO_2 is absorbed. Determine how this principle can be used in the design of experimental apparatus for studies involving living organisms.

Demonstrations Involving Student Volunteers

1. *Demonstration of CO_2 production during the process of fermentation:* In a 500 ml. Ehrlenmeyer flask, combine 0.5 g. Bacto yeast extract, 0.4 g. monobasic potassium phosphate, 8.0 g. glucose, 1.0 g. peptone, and 200 ml. distilled water. Heat the mixture gently to 37°C and inoculate with ½ envelope of dry yeast. Stopper the flask with one-hole rubber stopper fitted with a 2-inch glass tube, to which is attached a 12- to 14-inch length of rubber or flexible plastic tubing extended into 10 ml. of phenol red indicator solution in a nearby test tube (A) supported in a test tube rack. Place a similar phenol red indicator tube (B), not in contact with the fermentation flask, next to tube (A). Place the entire assembly in a convenient position for student viewing at 4 to 5 minute intervals. Meantime, review the properties of phenol red as an indicator and ask a student volunteer to bubble CO_2 through a straw into a fresh tube (C) of the indicator solution. Allow students to examine the color change occurring in tube (A) and compare with that (C) prepared by the volunteer and with the control (B). Students should be able to determine the source of the CO_2 in tube (A) and the biochemistry involved in the sequence of events leading from CO_2 production to the formation of carbonic acid detected by the phenol red. Other products of fermentation, other organic substrates, and a comparison of aerobic and anaerobic respiratory processes are logical follow-up studies for individuals and small groups to pursue.

2. *Demonstration of an epidemic:* In advance of the demonstration, prepare 7 sterile nutrient agar plates, a broth culture of *Serratia marcescens*, and a fresh orange, washed with green soap and held in a plastic bag until needed. Enlist the aid of 4 student volunteers and ask student (A) to stroke the thumb and fingers of his right hand over the agar surface of plate #1, then to repeat the process with plate #2 after having washed his hands with green soap. Thoroughly swab the entire surface of the orange with a sterile applicator that has been dipped in the bacterial broth culture, and offer the orange to student (A), instructing him to

1. handle the orange, making good thumb and finger contact, as if for eating,

2. shake hands with student (B),
3. repeat the agar surface streaking activity, using plate #3, and
4. wash hands with green soap.

Student (B) will repeat (A)'s performance of shaking hands with (C), stroking agar plate #4, and washing hands when finished; (C) will follow suit, shaking hands with (D), stroking plate #5, and washing hands; and (D) will stroke plate #6 and wash hands. Incubate all agar plates (including #7, which is a control) for 24 hours at 37°C and examine for bacterial growth. A lively discussion can be expected to follow this visual demonstration of an epidemic.

Student-prepared Demonstrations

Effective and worthwhile student demonstrations often grow out of independent study projects. By remaining alert to the progress being made by individuals as they engage in these activities, you can identify appropriate topics and encourage concerned students to perfect the techniques necessary for demonstrating them to others. Not only does the demonstrator gain an advantage by reinforcing his understanding of the topic, but all students profit when the sharing of what has been learned is viewed as an important outcome of the learning experience. It should be remembered, also, that the purpose of a student demonstration—to facilitate learning for all—must be ensured and, to achieve this goal, a practice run with you in attendance is a necessary preliminary part of the demonstrator's presentation.

Topics from almost every phase of biological study are good sources for student demonstrations, and advanced and highly motivated students are capable of presenting both simple and sophisticated concepts in understandable terms to their peers. Ten topics that have proved effective for special student demonstrations are:

- Koch's postulates
- Winkler method for determination of dissolved O_2 in water
- Miller method for determination of dissolved O_2 in water
- Power's method for determination of CO_2 in water
- Emerson enhancement effect in photosynthesis
- Chromatographic separations of biological substances
- Electrophoretic separations of biological substances
- Metabolic rate determination in small organisms
- Extraction of DNA from a bacterial strain or a thymus gland
- Diffusion through a membrane

LONG-RANGE AND ON-GOING DEMONSTRATIONS

There are many on-going activities in which important biological principles and phenomena are demonstrated without involvement in an experimental approach. Regeneration of planaria, geotropic responses of seedling roots in agar, phototropistic responses of both seedlings and established house plants, propagation of plants from bulbs, tubers, and cuttings, a "balance-of-nature" in an aquarium, and transpiration occurring in a terrarium are to be observed in most biology classrooms. By becoming alert to opportunities for calling attention to these "demonstrations" which are an integral part of the atmosphere, you can skillfully employ natural rather than "setup" conditions with a minimum of extra effort on your part and to a greater advantage for enriching the learning experiences of your students.

Demonstrations are among the most effective of vehicles for guiding the thinking of all students along similar pathways to reach similar understandings about basic topics in biology. A good demonstration lends variety to the usual fare of learning activities, and students seem to profit most from those that are presented dramatically, providing moments of suspense and encouraging speculation about the outcomes.

6 Unifying the Disciplines—Relating Biology to Total Learning

To say that biology is "the study of life" is, in reality, an oversimplification of its true nature. Inherent in any consideration of the biological sciences are interrelations with other disciplines; we depend upon their employment for explanations and understandings of the many aspects of that phenomenon which is Life and for gaining insights into the immediate and long range impact and significance brought about by interactions of living things with each other and with their environment. It is an all-encompassing study in which there are implications for incorporating elements of mathematics, the chemical and physical sciences, and both oral and written expression and communications. And it is through these interactions with other disciplines that students come to the realization that all of their learning is in some way related and that, when such application is made, there is (1) a greater mastery and/or reinforcement of the skill and knowledge being applied, and (2) a better understanding and perspective of the biological topic being studied.

MATHEMATICS IN BIOLOGICAL STUDIES

There are many biological concepts whose understandings are based on mathematical principles; total familiarity with these principles both enhances the student's learning experience in biology and reinforces his understanding of the mathematics involved. Through the interaction he is helped to grasp the meaning of the new concept and to work toward the attainment of a primary goal of education: the convergence of all aspects of learning in a single entity—himself—a totally educated individual.

Often referred to as a "tool" of the scientist, mathematics should be presented as an assist to the biologist in his investigative work. Insofar as possible, skills being taught in mathematics classes should be determined and correlated, where feasible, with studies in biology. The math learned in a math class invariably takes on a new importance and meaning when applied to a specific biology investigation in which the student is involved; and in this new context there is often a greater motivation for its mastery. However, with current trends toward flexible scheduling and open enrollment in classes, academic backgrounds of students tend to be highly diversified and represent a wide variety of levels of exposure, skill, and proficiency in math. Consequently, much of this implementation must be done on an individual basis, and you must be prepared to: accept the responsibility for introducing, developing, and applying appropriate mathematical procedures; be adept in teaching techniques for the use of formulas, calculations, and interpretations relating to mathematical data; and be innovative in presenting the application of mathematical procedures to biological studies. In their studies, students must be encouraged to gain self-confidence through the use of their mathematics "tool" with skill and dexterity so that the mathematics involved really is a tool and does not displace the main thrust of the biological concept that is being developed.

Doubling of Cells in Cell Division

It is helpful to students if an organized format can be devised that will assist them in the development and employment of the mathematical concepts they are to apply to their biological studies. For example, comprehension of the magnitude of numbers expressed in exponential form is essential to an understanding of many studies. In the case of doubling, a common process involved in successive cell divisions, students can learn to use exponents of 2 easily and with understanding if the concept development begins with an original cell doubling to form two cells. The second doubling is presented as 2^2 (or two 2s multiplied together = 4) cells, and subsequent doublings as:

$$2^3 = 8 \text{ cells}$$
$$2^4 = 16 \text{ cells}$$
$$2^5 = 32 \text{ cells, etc.}$$

Using this scale, students can become quite adept not only with the mechanics involved but with the comprehension of the magnitude of numbers as they are expressed in exponential form. In a bacterial population in which doubling occurs every 20 minutes under ideal and unrestricted conditions, this is translated in terms of 1 organism undergoing 3 doublings in 1 hour to produce 2^3 organisms, and 72 doublings to produce 2^{72} organisms in a 24-hour period of time. Students should research the generation time of cells of various species at various times in their lives and the effects of some internal and external environmental factors on the normal generation rates and patterns. Consideration should also be given to the consequences of the doubling time of fibroblasts (2 days in one instance) as compared to the longer (2.1 days) time for normal cells of the same species. Viewed from the perspective of mathematics, students can be motivated to consider important and significant biology-related issues with greater meaning and understanding.

DNA Coding for Amino Acids

This concept can also be applied to other situations using other number bases. In the case of DNA coding it becomes necessary to determine how the 4 distinguishable DNA nucleotides can be arranged in order to satisfy the need for a code that is specific for each of the 20-22 amino acids. Rather than have students engage in the time-consuming activity of preparing lengthy lists that spell out all of the possible combinations of nucleotides, they may be shown that:

- 4 things arranged 2 at a time will yield 4^2 different combinations
- 4 things arranged 3 at a time will yield 4^3 different combinations
- 4 things arranged 4 at a time will yield 4^4 different combinations

They should also be allowed to make a determination as to the number and the nature of base combinations needed for a workable code in this situation.

Determining Population Densities

The use of base 10 exponents with both positive and negative values offers convenience and ease of handling for very large and very small numbers involved in population studies and sampling techniques. One ml. of a 24-hour broth culture of *Escherichia coli*, for instance, might be said to contain 6,700,000 or 6.7×10^6 organisms. Students, of course, are aware that it is not possible to actually count this many organisms, no matter the form in which their number is expressed. Demonstrating a technique that employs serial dilutions and negative exponential expressions (see Figure 6-1) of the concentration levels offers them a convenient means by which to make such a determination. Using 1 ml. of a concentrated food dye in 9 ml. of water dilutes the dye to 1 part in 10, or 1/10, with further ten-fold dilutions producing 1/100,

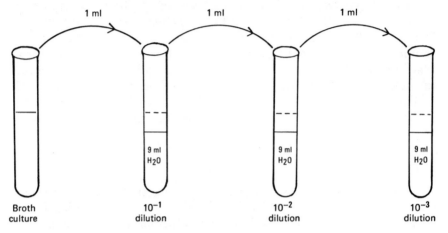

Fig. 6-1 Negative exponents may be used to express the concentration level of a serially-diluted culture

1/1000, . . . 1/1,000,000 dilutions of the dye. A visual interpretation of the employment of negative exponents to the ten-fold serial dilution can be charted:

1/10	0.1	10^{-1}
1/100	0.01	10^{-2}
1/1000	0.001	10^{-3}
1/10,000	0.0001	10^{-4}
1/100,000	0.00001	10^{-5}
1/1,000,000	0.000001	10^{-6}

Thus, the bacterial culture cited, serially diluted and plated in 1 ml. samples of each dilution, might have produced a countable number of 67 colonies (representing 67 well-spaced organisms before growth into 67 discrete colonies) in its 10^{-5} dilution. Correcting for the dilution factor, 67 colonies in a 10^{-5} dilution is easily seen to be 67 x 10^5 or 6.7 x 10^6 or 6,700,000 organisms/ml. of the original culture. Taking another example, 13 colonies counted in the plating of a 1 ml. sample of a 10^{-6} dilution of another culture would indicate 13 x 10^6 or 13,000,000 organisms/ml. of the original culture.

UNIFYING THE DISCIPLINES

For conversion of base 10 exponential expressions to more commonly used number forms, students need only adjust (1) the number of zeros to correspond with the magnitude of the exponent, and (2) the placement of the decimal as indicated by its sign—to the right for a positive exponent, since 10 with a positive exponent is a number larger than 10, and to the left for a negative exponent, since this has a smaller value than the base number 10. Do not encourage students to remember these two steps as a rule to be followed. Instead, encourage them to think and to apply reason to the larger or smaller value of the absolute number as indicated by the sign of its exponent.

Calculating Blood Counts

Blood counts also employ calculations in which corrections for a dilution factor play a part and in which you can guide your students toward speedy computations via skill in the application of their mathematics tool. Since making a blood count is a highly personalized activity, it has a built-in motivation for students to work with care and bring some real meaning to the experience.

The standard procedure for counting erythrocytes specifies that a fresh blood sample be mixed with physiological saline solution in a blood diluting pipette to effect a concentration of 1/200. The transfer of 1 drop of diluted blood to the surface of a Levy-Hausser corpuscle counting chamber permits viewing, under 100X of the compound microscope, of erythrocytes spread over the central 1 sq. mm. ruled area, which is further divided into 400 smaller squares. Then, with 430X viewing, a count of all corpuscles appearing in the 4 corner and 1 centrally located squares (each containing 16 small units) will yield a total of 80 small squares counted. Considering the distance between slide and cover slip to be 0.1 mm., sufficient information is now available for calculating the number of erythrocytes per cmm. of blood according to the formula

$$C = \frac{\text{Total number of erythrocytes counted} \times \text{dilution} \times 4000}{\text{Number of squares counted}}$$

Example: If a total of 420 erythrocytes were counted,

$$C = \frac{420 \times 200 \times 4000}{80}$$

becomes an easy calculation if you call students' attention to the product of the two constants in the numerator (800,000) being divisible by the constant number of small squares counted (80) to give a value of 10,000. Since these are all constants in any application of the formula, it becomes necessary only to multiply the total number of cells counted in the 80 squares by 10,000, or to add four zeros, thus moving the decimal 4 places to the right to calculate the number of erythrocytes/cmm. of blood.

$$\frac{420 \times 200 \times 4000}{80} = 4{,}200{,}000 \text{ erythrocytes/cmm. of blood}$$

In variations of the technique and calculations employed, white blood cell counts, bacterial population counts, and yeast culture cell counts can be made in a similar manner.

Probability in Genetics Studies

Biology does not always deal with certainties. Frequently, it is concerned with what is most likely to happen in a given situation or with the probability of the occurrence of a specific event. In genetics studies, for example, we are concerned largely with what is most likely to happen when genes are passed from one generation to the next. If, for instance, we cultivate wild type fruitflies and examine their progeny, we find about equal numbers of males and females among their offspring. It appears that there is a 50-50 chance that each egg will be fertilized by an X-bearing (determiner of a female offspring) or a Y-bearing (determiner of a male) sperm cell, so that the probability that an offspring will be a female is ½, and similarly, for a male it is also ½.

To determine the probability of specific combinations of males and females in a sample of the progeny, however, further mathematical application becomes necessary. It is important to point out that when two or more flies are selected together in a population sampling, each is independent and has an equal chance of being male or female. Further, the ratio of the results obtained from a series of samplings will vary with the size of the sample, but the expected results can be predicted in accordance with the law of probability which states that "the chance of two or more independent events occurring together is the product of their chances of occurring separately." So the probability of a sample of two flies representing both males or both females is easily seen to be ¼, but since there are actually two ways that one of each sex can be represented in the sample (either of them can be male, the other female) the probability of this combination is 2(½)(½) or ½.

The probability of male and female combinations in larger samples can be determined by using an expansion of the binomial, usually expressed as $(a + b)^n$ where:

a = the probability of a male
b = the probability of a female
n = the size of the sample

UNIFYING THE DISCIPLINES

Most students have a familiarity with the binomial theorem and algebraic expansion of the binomial:

$(a + b)^2 = a^2 + 2ab + b^2$
$(a + b)^3 = a^3 + 3a^2b + 3ab^2 + b^3$
$(a + b)^4 = a^4 + 4a^3b + 6a^2b^2 + 4ab^3 + b^4$
$(a + b)^5 = a^5 + 5a^4b + 10a^3b^2 + 10a^2b^3 + 5ab^4 + b^5$
$(a + b)^6 = a^6 + 6a^5b + 15a^4b^2 + 20a^3b^3 + 15a^2b^4 + 6ab^5 + b^6$.

They recognize the decending order for exponents of a, from n in the first term, by 1 in each successive term, decreasing until it becomes 0 in the last. The exponent of b, conversely, is 0 in the first term, is increased by 1 in each successive term, and reaches n in the last. Thus, there are always (n + 1) terms in the expansion of $(a + b)^n$.

The coefficients of a binomial raised to any power can be represented geometrically in a two dimensional structure known as Pascal's Triangle (see Figure 6-2), where the coefficient of any term may be determined by the sum of the two above to its right and left.

Pascal's Triangle

												n
$(a+b)^1 =$					1	1						2^1
$(a+b)^2 =$					1	2	1					2^2
$(a+b)^3 =$				1	3	3	1					2^3
$(a+b)^4 =$			1	4	6	4	1					2^4
$(a+b)^5 =$			1	5	10	10	5	1				2^5
$(a+b)^6 =$		1	6	15	20	15	6	1				2^6
$(a+b)^7 =$		1	7	21	35	35	21	7	1			2^7
$(a+b)^8 =$	1	8	28	56	70	56	28	8	1			2^8
$(a+b)^9 =$	1	9	36	84	126	126	84	36	9	1		2^9
$(a+b)^{10} =$	1	10	45	120	210	252	210	120	45	10	1	2^{10}

Fig. 6-2 Pascal's Triangle for determination of coefficients of terms in the expansion of binomial (a + b) with exponents 1 - 10. (n = the sum of the combinations in a given expansion)

Thus, students have a means by which to derive an expression for combinations of males and females in a given population sample, and to apply it in other considerations involving probability in genetics studies. Of particular interest to students, especially if some work in a local hospital, is the probability that the babies born on any given day will be all girls, all boys, or any of the possible combinations comprising the total number.

Example: To determine the probability that of 5 babies born at a given hospital on the same day 3 will be girls and 2 will be boys, they need only use the appropriate term ($10a^2b^3$) of the expansion of (a + b) that provides for a sample of 5 to find that

$$P = 10a^2b^3$$
$$P = 10(½)^2(½)^3$$
$$P = 10/32 \text{ or } 5/16$$

Probabilities of other combinations can be determined, using appropriate terms of the expansion, and the most and least probable combinations identified.

When the probability of only a certain combination in a given size sample is desired, students may be shown how to apply the probability formula where:

n = total number in sample
x = number in one class
(n-x) = number in other class
p = probability of single occurrence of first class event
q = probability of single occurrence of other class event,

$$P = \frac{n!}{x!\,(n-x)!} \cdot p^x q^{(n-x)}$$

In the situation described above,

$$P = \frac{5!}{2!\,(5-2)!} \cdot (½)^2(½)^{(5-2)}$$

$$P = \frac{5 \cdot 4 \cdot 3 \cdot 2 \cdot 1}{2 \cdot 1\,(3 \cdot 2 \cdot 1)} \cdot (½)^2(½)^3$$

$$P = 10/32 \text{ or } 5/16$$

When analyzing data collected from their genetic studies with fruitflies, laboratory mice, or genetic plants, students should be encouraged to compare actual results with those expected according to genetic patterns. It is helpful if they design data charts in advance so that tabulations can be made in organized fashion and relationships can be seen. For example, in a *Drosophila* monohybrid cross:

UNIFYING THE DISCIPLINES

WILD TYPE × vestigial wing

P = WW × vv
F₁ = all Wv (mated Wv × Wv)
F₂ = WW Wv Wv vv
(expected phenotypic ratio 3:1)

Sample of 200 F₂ individuals

	(e) Actual	Expected	(d) Deviation
W—	158	150	8
vv	42	50	8

Testing Data by Application of Standard Deviation Rule

Students can be shown an application of methods of statistical testing for the determination of the magnitude of the deviation said to be *standard* for such a situation and the extent to which actual results may deviate from those expected and still be considered to be attributable to chance alone. Applying the formula for standard deviation (S.D.):

p = expected number of one class
q = expected number of other class
n = size of sample

$$S.D. = \sqrt{\frac{p \cdot q}{n}}$$

$$S.D. = \sqrt{\frac{150 \cdot 50}{200}}$$

$$S.D. = \sqrt{37.5} = \pm 6.1$$

Comparing the actual deviation (d) with what is determined to be standard (S.D.) for this situation, we find that:

$$\frac{d}{S.D.} = \frac{8}{6.1} = 1.3$$

Since twice the standard deviation is taken to be within the realm of chance deviation, these data appear to be acceptable, indicating that there is no apparent significant factor operating to account for the relatively small difference between the actual and the expected results.

Applying the Chi Square Method

Students will need another device for evaluating data where more than two classes are involved. In a classical dihybrid cross in which four phenotypes are represented in an expected ratio of 9:3:3:1 in the F₂ generation, students should learn how to apply the Chi Square test to data collected in an actual genetics study. Example: Fruitflies, homozygous for two traits were crossed:

NORMAL STRAIGHT WING   vestigial wing
       ebony body   ×   NORMAL GRAY BODY

P = SSee × vvGG
F_1 = all SvGe
F_2 = S—G—   S—ee   vvG—   vvee
(expected phenotypic ratio, 9:3:3:1)

	Actual	(e) Expected	(d) Deviation
S—G—	286	270	16
S—ee	97	90	7
vvG—	78	90	12
vvee	19	30	11

Applying the Chi Square formula,

$$X^2 = \sum \frac{(d)^2}{e}$$

$$X^2 = \frac{(16)^2}{270} + \frac{(7)^2}{90} + \frac{(12)^2}{90} + \frac{(11)^2}{30}$$

$$X^2 = \frac{256}{270} + \frac{49}{90} + \frac{144}{90} + \frac{121}{30}$$

$$X^2 = .95 + .54 + 1.60 + 4.03 = 7.12$$

enables students to compute the Chi Square value for the data. They should then be helped to recognize that for any given fruitfly appearing in a 4-class population, there are 3 degrees of freedom for its phenotype. Then, reference should be made to a Chi Square table:

Probability (P) for Chi Square Values
and Degrees of Freedom (d/f) 1-3

P d/f	0.99	0.95	0.80	0.70	0.50	0.30	0.20	0.05	0.01
1	0.000	0.004	0.064	0.148	0.455	1.074	1.642	3.841	6.635
2	0.020	0.103	0.446	0.713	1.386	2.408	3.219	5.991	9.210
3	0.115	0.352	1.005	1.424	2.366	3.665	4.642	7.815	11.345

Students should read from the table that for 3 degrees of freedom the probability of a Chi Square value of 7.12 falls between .05 and .20, which can be interpreted to mean that such deviations or greater than those observed in the actual sample of fruitflies may be expected to occur solely due to chance in 5-20% of similar samples of the same population if numerous samples were taken from the fruitfly population. It would not, then, be unusual in a sample of 480 F_2 fruitflies resulting from the cross made to find 286 expressing both dominant genes, 97 expressing the dominant gene for wing structure and the recessive gene for body color, 78 expressing the recessive gene for wing structure and the dominant gene for body color, and 19 expressing both recessive genes.

Mathematics as a Tool of the Biologist

Most biology students will have had some exposure to the mathematics involved in their biological studies, and even those who have not previously mastered the skills for handling square root, exponents, factorials, and binomial expansions can learn and apply the appropriate operations with meaning if given a proper incentive. Presenting them as tools needed for conducting some biological studies puts them in the same category as a microscope, a weighing balance, a centrifuge, or any other helpful device and seems to provide this incentive. Since taking this approach and helping students to develop this attitude, my colleagues and I have noticed a remarkable improvement in test scores for math-related biology topics and a growing popularity in the math-oriented genetics studies. Students have become more active in seeking extra help, when needed, from both biology and math teachers. And, when given specific hints on how to handle each type of computation, they have developed a greater self-confidence. No longer faced with the frustration of not knowing how to handle the math involved, they have been able to set a first priority on the main thrust of the biological study and to view in proper perspective the interrelationships that exist between the two disciplines.

BIOCHEMICAL AND BIOPHYSICAL PATTERNS

Via their biological investigations, students are made aware of the presence of various combinations of the same basic 20-22 amino acids in all protein substances, a similarity in the taxic and tropistic responses exhibited by diverse living forms, and the same mode of action in the decomposition of hydrogen peroxide—whether in the presence of certain bacterial species, fresh raw liver, or an open cut on a finger. These are among the many observations that make clear the fact that there are fundamental biochemical and biophysical patterns which pervade all biological systems: all organisms depend upon the same basic building materials for their structure; all derive energy from the same basic sources; all engage in the same physiological activities; all are influenced and governed by the same basic natural laws; and all are capable of responding to the same basic stimuli.

Physiology of a Blood Clot

Living organisms are generally viewed as well-organized biochemical systems, and the role of enzymes in respiration, fermentation, photosynthesis, and digestion of food have been well-documented in student investigative activities. However, personal experience indicates that student interests are heightened when additional life processes to which they can relate are also introduced. The field of human physiology is rich in opportunities for such application. Take, for example, the everyday experiences of a student in which he may suffer a cut finger while using a scalpel, a jabbed hand from a compass point, or a scraped knee or elbow in a sporting event accident. In each, there may be a minor loss of blood and the physiology of a blood clot becomes a meaningful activity. While viewing a drop of freshly drawn blood, at 100X on a microscope slide, students should focus on the edge of the drop and observe, within seconds, thin dark fibrin threads beginning at the edge and growing inward. In this regard, so as not to miss the initial stages, it is helpful if students first focus on a wax pencil line drawn on the surface of the microscope slide. Then, without changing the focus setting, a drop of blood can be placed on the slide so that when it is returned to the microscope stage it will be automatically in focus for immediate viewing. They should be encouraged to watch closely, over a period of 5-15 minutes, the building of a complete fibrin network—a blood clot.

An investigation into the physiology of the clot formation will provide some insight into the intricate and delicately balanced interaction of a number of chemical and physical factors: the role of prothrombin in initiating the formation of a clot only when blood is shed and/or tissue is injured; the importance of vitamin K and calcium ions; the enzyme action of thrombin in the conversion of the plasma protein fibrinogen to insoluble fibrin threads which become intertwined to form a meshwork; the role of platelets in the production of thromboplastin; and the physical entrapment of red corpuscles during the formation of a barricade across the wound opening to stop the flow of blood—all these are critical and can be summarized in a flow diagram as shown in Figure 6-3.

Many physical factors also affect the clotting process. Students should be encouraged to design some laboratory investigations which

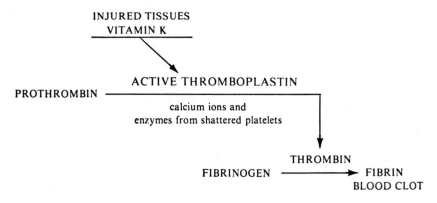

Fig. 6-3 Major events leading to the formation of a physiological blood clot

explore the effects of temperature and of rough wettable surfaces on clotting time. In this regard, they should reflect on some personal experiences and observations and make determinations concerning the practice of keeping blood samples under refrigeration and of applying wet gauze over an open wound.

Other situations involving the effects of varying concentrations of calcium dissolved in the blood, the use of anti-coagulants, and the nature and consequences of a thrombosis or of haemophilia should be considered in light of the study.

Mode of Action of Penicillin

The dynamics of osmosis are most dramatically demonstrated when students have the opportunity to view living cells immersed in hypotonic, hypertonic, and isotonic solutions. Red blood corpuscles, for example, can be observed to swell and burst when placed in distilled water, to shrink and shrivel in a 2% solution of NaCl, and to remain relatively unchanged in size and shape when in contact with a 0.85% physiological saline solution. This experience provides them with an understanding of why freshly excised organs and *Drosophila* larvae should be bathed in a Ringer's solution and why distilled water should never be used for maintaining hydra, planaria, or other invertebrates in the laboratory.

Observation of Elodea cells, however, permits students to view the rigidity of cell walls and their effects. As water enters the plant cells from a hypotonic solution, the cell content can be seen to push against the cell wall, building up a high internal pressure (often several atmospheres) until no more water can enter. Expanding this concept, students should identify the role of the cell wall and the mechanisms whereby conditions of turgidity and flaccidity are produced. They should recognize that the removal of a plant cell's wall would render it helpless in maintaining its rigid shape and, therefore, vulnerable to the effects of a hypotonic solution. They should also be helped to recognize that it is upon this principle that the mode of action of *penicillin* is based: by inhibiting the synthesis of muramic acid (an essential component of the cell wall of gram-positive bacteria), penicillin renders newly-formed sensitive bacteria vulnerable and, if not protected osmotically, destined to burst and be destroyed.

Students should be encouraged to consider the biophysical and biochemical implications here and to investigate other modes of action upon which the effectiveness and specificity of other antibiotics depend.

Effect of Scale on Organisms

The concept that an increase in size is a factor which brings about qualitative as well as quantitative changes in a biological system is often difficult for students to visualize. Equally effective on a cell, organism, population, society, or community level, a shift in scale does make a difference and may have some far-reaching effects. A change in scale may be seen to affect the surface to volume ratio of a cell or of an organism.

Understanding the scale effect and surface to volume ratio is basic to an understanding of cell size and energy needs, and a change in scale may be seen to exert influence on the cell or organism affected. Considering the energy needs—therefore the metabolic rate—of an organism, students can cut agar cubes of various dimensions and make simple calculations for a comparative study of their surface to volume (S/V) ratios. Data relating to the hands-on activity can be organized in the form of a chart for an at-a-glance comparison:

Dimensions of Cube (in cm.)	Surface Area (in sq. cm.)	Volume (in cu. cm.)	S/V
1 x 1 x 1	6	1	6:1
2 x 2 x 2	24	8	3:1
10 x 10 x 10	600	1000	0.6:1

Recognizing the tendency of all homoiotherms to lose body heat at the surface, it can be seen that there is a larger heat energy loss per unit of volume in the smallest "organism," and a relatively small loss per unit of volume in the largest. The solution then must come from compensating for the loss of energy by varying the metabolic rate, so that a mouse is observed to have a faster rate of metabolism than that of an elephant. Other similar comparisons can be made to assist students in their recognition of how living things have solved this basic problem: that by adjusting its metabolic rate an organism answers its acceptance of the consequences of the scale effect. This, of course, is somewhat altered by the irregular shape of a multicellular animal, and the effective shape is determined, at least in part, by physical and mathematical principles of which the S/V ratio appears to be all-pervasive in its influence.

Students should be encouraged to express ratios in simplest form, and to justify, on the basis of S/V, the microscopic size of most cells and the existence of some exceptions, such as the *Acetabularia* which sometimes attains a length of 7-10 centimeters.

SYNCHRONIZING BIOLOGY STUDIES WITH OTHER CLASSES

Other interrelations between biological studies and the chemical and physical sciences can be developed for individuals, groups, or entire classes. Some topics that lend themselves to this development are also very popular among students who engage in independent study projects, and some suggest demonstrations that can be performed by students:

Biochemical-Biophysical Topics

Differential Staining Principles and Procedures

Enzymes and Their Action

Biochemical Testing Procedures

pH and Life

Bioelectricity and Muscle Contraction

Factors Affecting Respiratory Rates

Energy Cycles in Fermentation, Respiration, and Photosynthesis

Reversible Reactions in Biosynthesis and Hydrolysis

Chromatographic and Electrophoretic Analyses of Biological Substances

Bioluminescence

Individual and group investigations into these topics can be arranged by keeping communications open with students and with other teachers, and by synchronizing schedules, where possible, to reinforce learning of biological studies via enrichment experiences. The approach is consistent with the concept of an integrated science; crossing over what are really artificial boundaries enables biology students to view living systems more realistically and with better understanding of the complexities of what is *Life*.

SOCIO-BIOLOGICAL TOPICS

Ecological, environmental, health-oriented, and population-related studies have implications for the social as well as the natural sciences. Not only do students need to have an understanding of the biological concepts of these many-faceted topics, but they also need the additional perspective of the structures within which they operate and interact.

Many schools have found a workable plan for satisfying these needs in interdisciplinary team-teaching programs where the emphasis is clearly placed on integration of knowledge in a unified course of study. Not limited by narrow confines and artificial boundaries of separate disciplines, the team approach helps students to develop an awareness and understanding of the biological issues that may affect their lives, and of the immediate and long-range effects they may also have on society. Even without the leadership and organization of a formal "team" structure, some innovative biology teachers are employing an interdisciplinary approach to appropriate topics for study. Some have found the answer in minicourses, some make extensive use of resource persons and guest lecturers, and most initiate reciprocal arrangements between themselves and a social science teacher in the interest of complementing each other's courses and enriching the learning experiences of their students through fitting the substance of what they teach to life's full fabric.

One school is experiencing considerable success through the implementation of companion courses—one directed toward the biological, the other toward the social aspects of a specific topic such as *population*—scheduled in sequential semesters. To ensure a correlated and unified study, the participating teachers work closely together, meeting frequently to compare notes and to determine the content and direction of the topic development. Both have freedom and flexibility in employing techniques according to individual preferences and expertise, and both have found judicious use of professionally prepared sound-slide and sound-filmstrip programs* to be helpful for presenting

---

*Excellent materials are available from Science and Mankind, Inc., White Plains, New York and The Human Relations Media Center, Pleasantville, New York.

relevant topics in an interesting fashion and for providing socio-biology learning experiences which are whole and coherent.

USING LIBRARY AND RESEARCH MATERIALS

Reading and communications skills are closely allied to biological study. Although today's approach unquestionably emphasizes inquiry and investigation, a student simply cannot set out in search of the answer to a solvable problem without first developing an orientation to biological science—an understanding of its basic laws, concepts, theories, postulates, and principles, and the practical aspects of their application. Library references and periodicals supply some of the background, and the use of multi-media resources and current literature, while supplementing and up-dating the basic text, lends a greater depth and perspective to a topic under consideration.

By establishing a good rapport and working relationship with the school librarian, you can help students to become familiar with the location, availability, and use of library materials. Jointly prepared bibliographies for special topics are of invaluable assistance, but students also profit from opportunities to browse, to explore, and to discover some materials on their own and to make selections according to their personal preferences. They should be encouraged to widen their horizons by experimenting with a variety of resource materials and to upgrade their reading levels as they indicate progress and readiness to do so.

There are many problems that are inherent in what students view as the *heavy* reading in biology. Skills developed for general reading are inadequate for handling technical references, and students may end up plodding their way through a maze of unfamiliar terms, expressions, and connotations. To avoid this pitfall, we must make biology an interesting and dynamic study that will give students reasons for wanting to seek information from the printed word. One teacher who travels over 250 miles each weekend to study a salt marsh environment reports that some of his interest and enthusiasm for the study rubs off on his students, as together they share some of the excitement of looking up information relating to the specimens he brings back to the classroom.

Readiness for dealing with technical literature requires the development of skills for competency in skimming and scanning and of organization for recall and critical appraisal. The aim should be the attainment of maximum comprehension with an economy of time and effort, using a format which is versatile and easy for students to follow. A basic outline with variations to accommodate text or outside sources, regular or advanced level students, and required or supplementary reading can be developed. Actual employment of such a plan has been amply rewarded in the form of improvement in reading rate and comprehension test scores and with the added bonus that students thus exposed seem to read more and not shy away from the "heavy stuff." Students also seem to gain considerable benefit from the use of a basic format for the reading of biological literature:

1. Survey the information to be read; read the title, sub-titles, skim the material for a general idea of its content, and examine the illustrations.
2. Establish a purpose for the reading: consider what the material has to offer—factual information, presentation of a new theory, etc.—and determine if it suits your purpose.
3. Preview any end-of-chapter or review questions to ascertain more precisely the nature of the content.
4. Read through the entire section to identify the main topic.
5. Re-read the entire section to determine the development of the main topic and its sub-topics.
6. Analyze and learn the meanings of new vocabulary terms; note familiar roots, prefixes, and word usage in context.
7. Re-read material again, according to established purpose.
8. Answer questions, if any, as a test of comprehension.
9. Evaluate answers—either by rapid feedback from teacher-corrected papers or student self-check.
10. Write a summary of what was read.

It must be remembered that good reading requires concentration and active mental participation on the part of students as they practice efficient skills. They must be given directed opportunities to organize their thoughts while reading, to practice the making of comparisons while reading, and to practice recall without re-reading. You can help them to deal successfully with their reading for biological study by emphasizing the importance of setting goals to be reached in each phase of their reading and by stressing three basic points recommended by most general reading improvement laboratories:

1. Several quick readings of the same material, with a well-defined purpose for each, are more fruitful than a single, slow, laborious one.
2. Reading the questions to be answered is as important as writing the answers. Remember, you cannot answer a question correctly unless you know specifically what is asked for.
3. Writing is the end product of reading. It is a fitting conclusion

for a reading assignment, be it a short paragraph or a long-range reading in a text, reference, or source book.

The importance of learning to read and follow directions with care cannot be overemphasized, particularly with students for whom biology is a first laboratory experience. Some chemical might be potentially dangerous if misused and, often, the success of an investigative study depends upon careful attention to directions as given. To test student ability to read and follow directions carefully, I use a variation of a test (see Figure 6-4) which is widely used with employees in business and

Name _____

Date _____

TEST ON FOLLOWING DIRECTIONS

1. Read carefully all of the following directions before doing anything.
2. Print your name, last name first, on the top line following the word "name."
3. Draw a circle around the word "all" in direction number 1.
4. Underline the word "name" in direction number 2.
5. In direction number 4, draw a circle around the word "underline," and in sentence number 1, cross out the word "anything."
6. Draw a circle around the title of this paper.
7. Circle the numbers of sentences numbers 1, 2, 3, 4, and 5, and place an X over number 6.
8. In sentence number 7, circle the even numbers and underline the odd numbers.
9. Write "I can follow directions" above the title of this paper.
10. Underline the sentence you have just written.
11. Draw a square about ½ inch to the side of the upper left hand corner of this paper. Draw a circle around the square.
12. Cross the numbers 8 through 12. Now circle the same numbers.
13. Put an X in the square inside the circle in the upper left hand corner.
14. In the space under the last direction of this paper, copy neatly, in your best handwriting, direction number 1.
15. Now that you have read all of the directions as stated in direction number 1, follow direction number 2 and number 16 only. Do not follow any other directions.
16. Please do not give away this test by any comment, exclamation, or action. If you have read this far, just pretend that you are still writing. It will be fun to see how many people really do follow directions carefully.

Fig. 6-4 Test used to reveal to students how carefully they read and follow directions

industry for the same purpose. It is a dramatic way to illustrate that many people really do not read and follow directions as carefully as they think they do and to give them some insight into the reasons why things sometimes go wrong. After the test has been completed, we discuss what consequences might ensue in the biology laboratory—and in other life situations—if this ability is not developed and practiced faithfully.

BIOLOGY AND GENERAL EDUCATION

Students must learn to read, to write coherently, to compute, to integrate their skills and knowledge of all interrelated disciplines, and to relate all of these experiences and competencies to life situations. But dull, tedious, rote methods and contrived situations should be avoided in favor of innovative and natural development. For example, familiarity with the Metric System develops best in a natural setting where its exclusive use—references to liter beakers, 10 ml. pipettes, 250 ml. flasks, 6 mm. cork bores, 65 g. laboratory mouse, 28°C. incubating temperature, and 4500 ccm. lung capacity—helps students to develop a concept of the magnitude of the measures rather than complicating the issue via application of conversion scales that somehow seem to perpetuate the less scientific measures and diminish the importance of the metric. In a further effort to *think metric* in the classroom, we may refer to the demonstration table as the 1 meter high table, to frequently used plastic rulers as 15 cm. rulers, to a star basketball player as the 2 meter tall star, and to a field trip to a facility visited as a 4.5 km. journey. One teacher, finding that her students had difficulty in visualizing the size of an acre of land referred to in population density studies, had her students work with students in an applied math class on a project to map out an acre on school grounds, using well-known landmarks as cornerstones. Any other size area could be similarly delineated, if needed, using existing markers—trees, corner of the building, etc.—as long as the area enclosed is the equivalent of the square measure desired.

Students learn best when that which is studied has, for them, true meaning. Whether interdisciplinary or related to everyday experiences, the usable knowledge becomes easier for students to learn and less difficult for teachers to teach, and it produces better results and fewer frustrations for both.

7
Social/Moral Issues in the Biology Classroom

The present rate at which biological science and its technology are growing is without precedent. Over the past few decades improved surgical procedures and organ transplants, sophisticated computerized blood analyses, more efficient methods for disease prevention, insect control, and food preservation, expanded employment of antibiotics in combatting infectious disease, more effective treatment and cures for some non-infectious diseases, and microprocessor pacemaker implants have become realities. In the offing appears to be the fulfillment of a promise for freedom from genetic disorders, the alleviation of some of the consequences of aging, expanded food sources, more effective methods for early detection and precise diagnosis of a wide spectrum of diseases, and implanted microprocessors for lessening pain, moving artificial limbs, and restoring sight to the blind and hearing to the deaf. Technological progress has enabled man to solve many age-old problems, and is continuing to do so; in its forward-looking approach it is aimed toward preparedness to cope with the future—toward the forestalling or prevention of problems projected for the years to come.

Indeed, man's technology has undergone change more rapidly than has his ability to change his attitudes to keep pace with it. For example, some of the methods involved in recombinant DNA research are viewed with apprehension and fear that the risks involved are inordinately high and that the whole practice of man's interference with natural design may get out of hand. Research workers themselves are looking to the NIH (National Institute of Health) to supply them with guidelines for this work. In another case, recommendations from the scientific sector for methods by which population can be brought into harmony with food, space, and energy resources are interpreted by some as threats to the rights of individuals and deemed to be incompatible with natural laws and social/moral ethics. Individuals and groups are speaking out and demonstrating against practices they consider to be unsafe, unwise, or in other ways contrary to their beliefs and attitudes; others are accepting and promoting, by word and action, the new technologies. There are many interactions between the scientific community, religious groups, the courts, government agencies, organized citizen groups, and individuals as various legal, social, moral, and ethical aspects of the proposed technologies and their implementation are being questioned and tested. In the framework of this set of checks and balances there are implications for individuals to have a voice in matters and to think and act in a responsible manner. They must become aware, informed, and concerned, and they must give recognition to the existence of a lag between technological change and individual and societal attitudes. Our aim should be to help students toward closing the gap through the making of adjustments, in either direction, as indicated by intelligent and informed thought and value judgment leading to responsible decision-making.

Clearly, students who participate in Right to Life marches and ecological crisis demonstrations are already shouldering grave responsibilities in this direction, and others, too, will be called upon to share in the decision-making and molding of mankind and his world of the future. This requires that the issues be faced with reality and that the processes of science be employed for problem solving based on analysis and interpretation of pertinent data, rational thinking, and consideration of alternative actions.

DEVELOPING SCIENTIFIC ATTITUDES

Opportunities to observe the attitudes and workings of biological science can be provided through the use of case histories involving such past issues as the *Cause of Malaria*, the *Opposition to Vaccination for Smallpox*, and *Biogenesis vs. Abiogenesis* in the controversy that raged over the issue of *Spontaneous Generation of Life*. Parallels can be drawn between the earlier experiences and our present day situation as a continuing story unfolds in the saga of the technological knowledge/attitude relationship. Exposure to case histories has been found to be especially helpful in encouraging students to recognize the importance of interaction between facts and ideas in science and in providing them with a historical perspective of the effects of new knowledge on the acceptance or rejection of a new attitude or point of view.

Demonstrated and practiced as a process of enquiry, biological science and its elements of investigation into relevant topics can be viewed by students in their proper relation within the dynamic system: through objective analysis and interpretation of pertinent data, man's knowledge grows and brings about changes in concepts which he uses in interpreting still other data, so that our knowledge is constantly growing and changing. Based on the best tested facts and concepts at any given point in time, the process of enquiry thus provides man with the most rational, but not necessarily absolute, knowledge. It is in this framework that scientific attitudes are developed.

Ways to Develop Skills for Interpreting Data

First and foremost, students must learn to analyze and interpret data accumulated by researchers in their experimental studies. We can help them develop this ability by introducing first level skills, such as those of reading graphs based on laboratory experiments or classical studies.

Example:

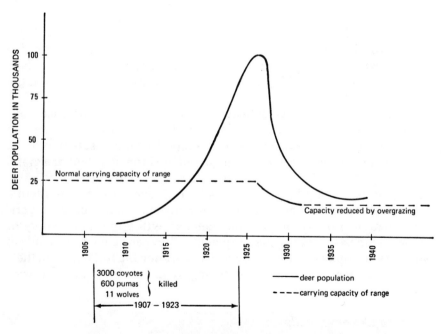

Fig. 7-1 History of the Kaibab Plateau in Arizona

Analyze the graph and indicate if the accompanying statements are:

A—True because the data supports the statement
B—False because the data refutes the statement
C—Not determined to be true or false by the data presented

- Prior to 1907, the deer herd was kept below the carrying capacity of the range by natural predators.
- Reducing the number of predators allowed the deer population to exceed the natural carrying capacity of the range for a period of more than 5 years.
- All individuals in excess of the normal carrying capacity of the range starved to death.
- Correction of the overpopulation condition restored the carrying capacity of the range to its original normal carrying capacity.
- Prior to 1907 neither disease nor starvation affected the Kaibab deer.

Higher level skills for data and graph analysis and interpretation can be developed in areas of student ability to understand the relevance of data to a problem under consideration and the ability to identify principles involved. Appropriate statements can be phrased for pertinent data for students to evaluate in terms of *logical or illogical hypotheses, interpretations consistent or inconsistent with the data, and a logical interpretation or a restatement of the data*. Or direct questions requiring student answers can be constructed to accompany a set of data presented. In all cases, student responses should be reviewed and discussed, with corrections made for any misinterpretations of the data. The aim should be to help them toward mastery and expertise in the skill.

Having developed analytical skills via guidance and practice, your students should now be ready to give attention to research reports appearing in scientific journals. If the report contains sufficient data, they should be encouraged to arrange it in the form of a chart or a graph for interpretation according to the pattern and skills developed. As a further assist in handling the information, students will find guideline questions (Figure 7-2) both helpful and welcome. It has been my experience that such questions, if made general, will apply to a variety of research reports. If duplicated in quantity, they can be made available to students on request for their guidance in deriving the full import of their specialized reading and for reducing their frustrations to a very minimum while in the process.

Ways to Develop Skills for Making Predictions

Students also need to have practice in making predictions. They should be encouraged to use available information concerning what has

GUIDELINE QUESTIONS FOR RESEARCH REPORT ANALYSIS

What was the problem being researched?

Is the problem one of significance and importance?

What facts were presented?

What assumptions were made?

What familiar techniques were employed in the research?

What new techniques were introduced?

Can the experiments performed be repeated with the same results?

What were the conclusions drawn by the experimenter?

What questions were left unanswered?

Are there any contradictions in view of previously or currently accepted ideas?

Are there any ambiguities in the line of reasoning presented?

What generalizations can be made on the basis of the study?

What effect will this study have on man and on society?

Fig. 7-2 Guideline questions help students in the preparation of a Research Report Analysis

been observed to happen in a given situation as a basis for making predictions for what might reasonably be expected to happen as a next step. It is a way of reinforcing their learning and of enabling them to develop a scientific attitude for approaching and interpreting experimental studies with thought and foresight.

Students find prediction-making very stimulating when it is presented in a challenging fashion. A technique which I have found to work successfully makes use of an overhead projector and transparencies, preferably with overlays and duplicated sheets of the original for distribution to students:

Example 1: As a follow-up study of DNA replication and RNA synthesis, project a transparency showing a segment of a strand of duplicating DNA and ask students to indicate on their copies the nucleotide sequence in each replicated strand. Then, allow students to check their responses with that of the overlay showing the correct sequence. They can identify errors, if any, and determine how, where, and why they were made. (A similar prediction of the nucleotide sequence in a strand of RNA synthesized by DNA can be made and checked.)

Example 2: Project a transparency showing diagrams representing the accumulated information from planarian regeneration experiments showing growth of whole animals from isolated heads, tails, middles, lefts, and rights. Also include a new situation—that of an otherwise whole planarian, with the head only split down the middle. Then ask students to make a diagram on their duplicate sheets, indicating their predictions about the appearance of that animal ten days hence. Again, comparisons of student responses with the overlay showing the results of an actual experience will serve as a reinforcement of student learning of the regeneration process and of their ability to apply experimental findings to a new situation.

This technique can be adapted to suit other situations based on accumulated data and trends indicated. Studies involving rates of growth of competing populations and tropistic responses in plants are particularly good sources of excellent material that can be handled effectively by students. In each case, the reinforcement of learning involved contributes to the development of skill in analyzing and interpreting data and assists the student in establishing an attitude of purpose, meaning, and application of the experimental data collected.

DETERMINATION OF WHAT IS LIFE

Respect and concern for *life* is dependent upon the recognition of what life really is and what distinguishes it from non-life. Students have no difficulty in citing examples, but their attempts at a definition do not indicate true understanding of its exact nature. Viewing a demonstration of a "mercury amoeba" or a "beating heart" invariably prompts a reexamination and refinement of preliminary definitions which reflect the sentiment that "if it moves, it's alive."

Demonstration: The MERCURY AMOEBA

Using an overhead projector, project the image of a watch glass resting on a white background. Partially fill the watch glass with dilute HNO_3 and introduce a drop of mercury to the pool. Then, using forceps, carefully place a large crystal of potassium dichromate about 1-1½ cm. from the mercury. Do not disturb the preparation, but allow students to view the activity: as the crystal dissolves in the acid it diffuses outward and, upon reaching the mercury, causes the mercury drop to flow in a manner similar to that observed in an active amoeba. Students should be asked to describe the activity and to determine if it indicates life.

Demonstration: The BEATING HEART

Before the demonstration, place a small pool (about 2 cm. in diameter) of mercury in a watch glass and cover it with 6M H_2SO_4. Add 1 ml. of 0.1M $KMnO_4$ and place a piece of iron wire, whose length is about equal to the radius of the watch glass, so that one tip touches the mercury and the other rests alongside the edge of the watch glass. Carefully place the entire assembly on the stage of an overhead projector and allow students to view as you slowly add 1 to 2 ml. of 16M H_2SO_4 until a rhythmic motion is observed to occur in the mercury. Students should observe the initiation and continuance of the motion which resembles that of a beating heart. They should be asked to describe the activity and to make a determination if the "heart" is alive.

Students should discuss the mercury amoeba and/or beating heart and express their opinions as to whether or not the objects viewed on the screen were alive, giving reasons, pro and con. After the discussion, you can reveal the chemical composition of the moving objects and encourage a consideration of whether any chemical or mechanical contrivances do, indeed, constitute living substance.

The demonstrations elicit great response from students, who, interacting with one another, are moved to give a critical appraisal of life and its characteristics. Research into legal and medical definitions should be reported, with emphasis on the specifics of when life begins, when it ends, and what man's attitudes toward its existence within this time interval are and should be.

Making a Determination:
When Did Sammy Die?

In a more innovative approach, one teacher has reported that when he introduces the study of life and students are trying to formulate a workable definition, he stimulates thinking about what life is and when it ceases, using the following story about "Substituted Sammy."*

> Sammy was a normal, healthy boy. There was nothing in his life to indicate that he was any different from anyone else. When he completed high school, he obtained a job in a factory, operating a

---

*Donald F. Shebesta, "Substituted Sammy: An Exercise in Defining Life," reprinted from The American Biology Teacher, Vol. 34, No. 5, May 1972; courtesy of The National Association of Biology Teachers, Inc., 11250 Roger Bacon Drive, Reston, Virginia 22090.

press. On this job he had an accident and lost his hand. It was replaced with an artificial hand that looked and operated almost like a real one.

Soon afterward, Sammy developed a severe intestinal difficulty, and a large portion of his lower intestine had to be removed. It was replaced with an elastic silicone tube.

Everything looked good for Sammy until he was involved in a serious car accident. Both of his legs and his good arm were crushed and had to be amputated. He also lost an ear. Artificial legs enabled Sammy to walk again, and an artificial arm replaced the real arm. Plastic surgery and the use of silicone plastic enabled doctors to rebuild the ear.

Over the next several years Sammy was plagued with internal disorders. First, he had to have an operation to remove his aorta and replace it with a synthetic vessel. Next, he developed a kidney malfunction, and the only way he could survive was to use a kidney dialysis machine. (No donor was found to give him a kidney transplant.) Later, his digestive system became cancerous and was removed. He received his nourishment intravenously. Finally, his heart failed. Luckily for Sammy, a donor heart was available, and he had a heart transplant.

It was now obvious that Sammy had become a medical phenomenon. He had artificial limbs. Nourishment was supplied to him through his veins; therefore he had no solid wastes. All waste material was removed by the kidney dialysis machine. The heart that pumped his blood to carry oxygen and food to his cells was not his original heart. But Sammy's transplanted heart began to fail. He was immediately placed on a heart-lung machine. This supplied oxygen and removed carbon dioxide from his blood, and it circulated blood through his body.

The doctors consulted bioengineers about Sammy. Because almost all of his life-sustaining functions were being carried on by machine, it might be possible to compress all of these machines into one mobile unit, which would be controlled by electrical impulses from the brain. This unit would be equipped with mechanical arms to enable him to perform manipulative tasks. A mechanism to create a flow of air over his vocal cords might enable him to speak. To do all this, they would have to amputate at the neck and attach his head to the machine, which would then supply all nutrients to his brain. Sammy consented, and the operation was successfully performed.

Sammy functioned well for a few years. However, slow deterioration of his brain cells was observed and was diagnosed as terminal. So the medical team that had developed around Sammy began to program his brain. A miniature computer was developed; it could be housed in a machine that was humanlike in appearance, movement, and mannerisms. As the computer was installed, Sammy's brain cells completely deteriorated. Sammy was once again able to leave the hospital with complete assurance that he would not return with biologic illnesses.

During the course of this presentation, Mr. Shebesta reports that he stops and asks the students if Sammy is still alive. The students are allowed to discuss and argue their viewpoints. Generally, all students agree that he is alive at the beginning of the narration and not alive at the end of it. But there is always some disagreement in identifying the point at which "Substituted Sammy" ceases to live.

Having used Mr. Shebesta's presentation, I can concur with his reported experience regarding student responses. Invariably, students want to be told when Sammy died. But Mr. Shebesta wisely refrains from giving answers. Instead, he points out strengths in student positions and viewpoints and advises that these be incorporated into student notebooks for reexamination and discussion throughout the development of the course. It is from these considerations that his students develop a definition of life and continue to examine and refine it as the course progresses.

SOCIAL/MORAL ISSUES IN BIOLOGICAL SCIENCE

Explorations into the medical and legal definitions of *life* and *death* are important also to considerations of abortion and the removal of organs for transplant, and to the existence of so-called "vegetables" in institutions. *Right-to-Die* and *Death-with-Dignity* also reflect implications for reconciling issues with social and moral codes, and specific cases, such as the well-publicized Karen Ann Quinlan case of a girl in a prolonged coma, offer good subject matter for consideration.

Refusal to accept medical, surgical, and amputation therapy deemed necessary to save a life have implications for the responsibilities of society, government, and the medical community, as well as for the rights of individuals. Conflicts which sometimes arise when medical personnel, dedicated and committed to the saving of lives, meet head-on with individuals who wish to be spared the prolonged agony and suffering and the depletion of financial resources brought on by futile attempts to deal with terminal cases by heroic measures should be researched, with full disclosure of the impact of the Hippocratic Oath, malpractice laws, religious beliefs, social and moral ethics, and the rights of individuals.

It is clear that opinions will differ as to how to deal with these situations. Indeed, there may not be one right answer for many of the issues about which there are controversies. Thus we must help each student to decide his own value position, as value judgments rather than absolute answers apply to most controversial issues. This can be seen to add to the dynamic nature of the study; via directed activities in examining, probing, researching, and considering the issues, student thought and experience can be directed toward the real life situations that need to be faced in a world whose society is making greater and greater demands as its technology continues to grow and become increasingly complex. Some timely controversial issues related to biological science include:

>Abortion
>Birth Control and Population Control
>Drugs and Drug Abuse
>Evolution
>Euthanasia
>Environmental Pollution
>Food Additives
>Cloning Practices
>Cryogenics
>Insecticides and Their Use
>Virus Pesticides
>Genetic Engineering and Recombinant DNA Research
>Right-to-Die and Death-with-Dignity
>Use of Experimental Animals in Research
>Sex-reversal Surgery

Techniques for Dealing with Controversial Issues

As with other controversial questions, those with social/moral overtones must be examined fully and considered from both positive and negative viewpoints.

INTERVIEWS with clergymen, doctors, lawyers, research biologists, psychologists, funeral directors, social workers, parents, students, and teachers will provide a wide sampling of values from different vantage points. If a panel discussion or other live appearance can be arranged, the advantage of a face-to-face dialogue with opportunities for questions from the audience will enhance the value of the interview, but, however accomplished—live, on tape, or reported by the interviewer— adequate preparation must be made: guest resource persons, interviewers, and audience must be supplied with guidelines; all must prepare by researching the topic, and all must be prepared to respect the opinions and viewpoints expressed by others.

DEBATES can be organized and arranged for almost any controversial issue. Two students to uphold the positive viewpoint, two to uphold the negative, and one to act as a judge must be selected. As determined by the length of the class period, specified equal time periods should be allotted to each speaker to present his case, with one-minute rebuttal time provided for each side. The judge should rule on what can be accepted as admissible, as determined by supporting evidence, and the remaining class members should decide on the winning team. It is required that all class members participate. The principals have specific roles to play, but all in the audience should research the topic being debated, judge the presentation by the supporting evidence, and practice tolerance for the "other side's" opinions. After the debate, convincing evidence presented by both sides should be recognized and discussed.

DISCUSSIONS provide one of the best ways to encourage all students to participate actively in a study. One successful technique is to allow them to first meet in small groups to research, plan, and discuss among themselves what position they will assume regarding a controversial topic and how they will support it. After the sub-group meetings, all students should assemble for an open discussion, sometimes with one member of each small group acting as its spokesperson. In this framework, often the discussion leader or the teacher must play the role of the Devil's advocate for the purpose of provoking responses and stimulating thinking, but always with respect and courtesy shown for all contributions made to the discussion. It is important that every student know that what he has to say will be listened to by others.

Students who for personal reasons hesitate to participate in an open discussion on either so-called or real moral principles can become contributing members in a different capacity. Given recognition for their impartial posture, they can be asked to prepare a short summary at the end of the discussion period, or they can be asked to conduct a follow-up study in the form of a student opinion poll or an interview with one or more persons of authority relative to the issue—and to report back to the group with their findings. Personal experience with these approaches has proved very satisfactory. When the technique has been appropriately matched to the individual and to the nature of the topic, contributions made by otherwise non-participating students have added significantly to the study; there is evidence to indicate that all students tend to develop a growing awareness of the importance of input by every other student as a means of clarifying issues and formulating individual positions, and the student originally identified as "reluctant" often becomes increasingly involved in a personal way as he makes his contributions to discussions throughout the term.

PLACING AN ISSUE ON TRIAL, using a simulated courtroom procedure, is a popular activity which involves many students as participants. The prosecutor in any given trial situation may be charged with the responsibility of presenting a case against a controversial substance (Laetrile, saccharin, marijuana), practice (insecticide spraying, use of food additives), or unauthorized program (self-prescribed medication or crash diet) which is potentially hazardous. The defense attorney may be defending an unauthorized product, an act of refusal of surgical amputation, or practice of an unlawful act under unusual circumstances. Each should build his case by calling witnesses to give testimony which the judge can rule upon, disallowing that which is not valid and that which cannot be properly documented or supported. On the basis of the evidence, members of the jury panel must decide the fate of the issue on trial. In a variation of this technique, the issue may be undergoing investigation by a senate committee, and testimony for and against the issue must be heard and evaluated by the committee members.

INDIVIDUAL ENRICHMENT EXPERIENCES can be provided for students in the form of conducting a full investigation into one particular phase of a topic of interest, about which there may be a controversy. By integrating laboratory and library research findings with information collected from recognized authorities in the field, they can then present their findings to the class for discussion. Many suitable topics for enrichment experiences are suggested by popular beliefs and advertising claims relating to foods, nutrition, health habits, and health care products. It has been my observation that high school students approach these studies with enthusiasm, mainly because they are topics to which they can relate and because there is much freedom in the manner in which the study can be pursued.

*Ways to Deal with Some
Controversial and Sensitive Issues*

1. POPULATION

There are controversies in abundance about proposals for control of the world population and about the means, if any, by which this should be accomplished. Many pertinent questions have been posed: Should the birth rate be limited to effect a zero population growth rate? What are the consequences of a continued increase in the rate at which our population is growing? Is the longer life span of individuals the major cause of the increased population figures? Has the earth reached its carrying capacity? Can the earth really accommodate, as one estimate indicates, 50,000,000,000 people? If there is no self-imposed control on population growth rates, will there be a natural control? If so, what form might it take and with what consequences? What limiting factors in the past have exerted an effect on man's population? How do famine, pollution, and over-crowding affect populations? Should man prepare now, slow the growth rate, and delay the time when the earth will no longer be able to support the life upon it? What are the alternatives?

In an attempt to understand the importance of these questions and to shed light on the over-all issue, your students can engage in a hands-on population study involving a bacterial population grown in the laboratory: a fresh nutrient growth medium is inoculated with non-pathogenic bacteria and 1 ml. samples are removed daily for counting by the conventional pouring of agar plates; then, from a physical count of the colonies grown from each discrete organism, the number of bacteria determined to inhabit a unit volume of the culture is observed to follow a characteristic growth pattern which can be divided into four main stages (see Figure 7-3).

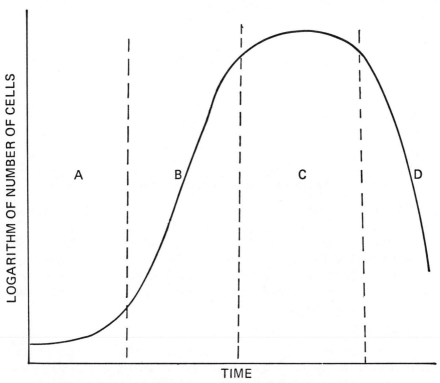

Fig. 7-3 Bacterial Population Growth Pattern showing A - Lag Phase, B - Logarithmic (Exponential) Growth Phase, C - Maximum Stationary Phase, and D - Death Phase

Students should then research and examine records of the human population growth from the year 4000 B.C. to the present, and examine its pattern (see Figure 7-4).

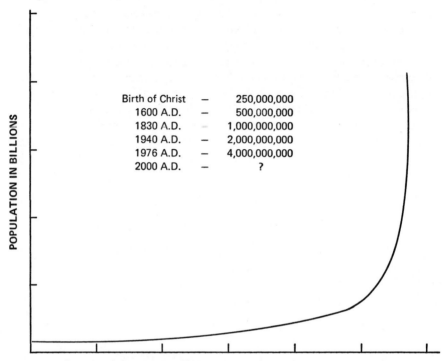

Fig. 7-4 Human Population Pattern, 4000 B.C. — 2000 A.D.

Patterns similar to the bacterial population growth can be seen to be emerging in the human population, and students should make determinations as to how long it will take to again double the human population, as to the stage of population growth man is now experiencing, and as to what the future portends. The study dramatizes the importance of the increasing rate of growth in man's population and the great concern about this issue.

2. SMOKING

The American Cancer Society has published some impressive reports indicating that the life span of cigarette smokers is shorter than that of those who do not smoke; the Surgeon General tells us that there is conclusive proof that smoking is injurious to health; and the number of cases of throat and lung cancer linked to smoking is increasing at alarming rates . . . yet the sales of cigarette companies continue to reach new highs.

To impress students with the enormity of the effects of smoking on individual smokers, Vivian and Albert Schatz have developed a dynamic activity for use in the Philadelphia public schools. They call it "One Puff of Death"* and begin by asking students "How much does one cigarette shorten a smoker's life?" The program then proceeds:

> To answer this question, you have to compare the average life span of smokers and non-smokers. You must also take into account the average number of cigarettes smoked each day. The American Cancer Society publishes the information you need to do this, for men; it was obtained from a survey of over one million people in the United States. The statistics for men are shown in the accompanying table (shown in Figure 7-5).

Life Expectancy in Years for Men of Different Ages

		Smoker			
	Non-	Number of cigarettes smoked daily			
Age	smoker	1-9	10-19	20-39	40+
25	48.6	44.0	43.1	42.4	40.3
30	43.9	39.3	38.4	37.8	35.8
35	39.2	34.7	33.8	33.2	31.3
40	34.5	30.2	29.3	28.7	26.9
45	30.0	25.9	25.0	24.4	23.0
50	25.6	21.8	21.0	20.5	19.3
55	21.4	17.9	17.4	17.0	16.0
60	17.6	14.1	14.1	13.7	13.2
65	14.1	11.3	11.2	11.0	10.7

Fig. 7-5 Life expectancy in years for men of different ages

> The data in the table show that a 25-year-old non-smoker has an average of 48.6 more years to live. A man of the same age who smokes 40 cigarettes a day can expect to live 40.3 more years. The difference in life span of these two individuals is 8.3 years. This is equivalent to 4,363,000 minutes.
>
> You may assume that these men began smoking at the age of 15, as many people do. At age 25, therefore, they will already have been smoking for 10 years. Assume that the second man will smoke for the remaining 40.3 years of his life. This amounts to 10 + 40.3 =

---

*Reprinted from *The American Biology Teacher*, April 1972, with permission of the authors.

50.3 years, or 18,360 days. Since he smokes 40 cigarettes a day, he will smoke a total of 40 x 18,360 = 734,400 cigarettes during his life. Calculating,

$$\frac{4{,}363{,}000 \text{ minutes}}{734{,}400 \text{ cigarettes}} = 5.94 \text{ minutes per cigarette,}$$

which shows that every cigarette he smokes subtracts approximately six minutes from his life.

Now assume that the man puffs each cigarette 8 times, as most smokers do. He therefore loses 5.94/8 = 0.74 of a minute or about 45 seconds of his life for every puff he takes.

But there is more than this to the story. The two packs of cigarettes he buys daily cost him an average of about 80 cents. Thus, during the 50.3 years that he smokes he pays close to $15,000 for cigarettes. He pays this money to die 8.3 years younger.

Smoking is therefore a kind of death insurance for which the 25 year old male smoker pays $15,000. This pays off for him at the age of 65.3, when he dies 8.3 years sooner than if he had not smoked. He pays $15,000/50.3 = $300 a year for 50.3 years. Looked at another way, it costs him $15,000/8.3 = $1,800 for each year that he shortens his life span by smoking.

People who smoke get more for their money than just a shorter life. There are other benefits in addition to an earlier death. Those who smoke have poorer health generally. They are also more likely to be involved in automobile accidents. Women who smoke during pregnancy have babies that weigh less at birth. A woman who smokes is nearly twice as likely to lose her baby, compared with a non-smoker. Students who smoke do not do as well in their studies. There are also more cases of cancer, emphysema, heart disease, bronchitis, peptic ulcer, and sinusitis among smokers than among people who do not smoke.

Do you smoke? If you do, how much are you paying for your death insurance? What are you getting for your money? How much sooner will you die? Calculate this for yourself (or for a friend who is a smoker), using the data in the chart and see how you come out.

The calculations involved in this activity are simple yet meaningful; through them students are able to recognize how they are personally affected by the health hazards of smoking. Keeping current is, of course, important, and you should check frequently the cost of cigarettes for use in calculating the premiums for a smoker's "death insurance."

3. HUMAN REPRODUCTION

Students are often reluctant to ask questions or to discuss openly any aspects of human reproduction. The solution to setting them at ease and respecting their sensitivities has been found by one teacher to lie in the use of a question box. Several days in advance of the session which she calls "all you ever wanted to know about reproduction but never asked," she provides a question box for students to write out their questions and submit, unsigned. She then organizes the questions into categories and prepares for the session. Equipped with a human torso model, overhead transparencies, and anatomical charts to illustrate, where indicated, she answers the questions honestly and objectively. Sometimes, using a clinical approach, she invites the school nurse or doctor to participate, or, if social, ethical, and moral aspects are indicated by the questions, she includes a guest clergyman, family counselor, or other qualified personage as well. The technique, however, is successful because it permits students to obtain accurate information without embarrassment and, as an extra bonus, she finds that her students rate this "session" as one of the most interesting and valuable in the course.

4. CONTROVERSIAL ISSUES, OLD AND NEW

Some of the controversial issues are relatively new. Cloning and genetic engineering, for example, pose social, moral, and ethical questions with legal implications for the thought of reproducing another Einstein, Hitler, or John Doe. Doubts exist as to whether man will create a future Utopia with genetically perfect specimens living longer, more useful lives, or a Doomsday disaster with a world inhabited by monsters. For our students, there are no answers. There are only unanswered questions and speculation, for we cannot foretell the future except in the hopeful predictions of researchers, in the fertile imaginations of science fiction writers, or in the long-range philosophies which encompass all.

Other issues are of long standing. Those issues once faced may need to be faced again, as in the case of the Evolution-Creation controversy which, in the minds of many, has never been fully resolved. Again, there are no absolute answers, for there are no documented proofs, and, as their names indicate, both views are classified as *theories*. Many teachers report that special care must be taken to ensure that students make a clear distinction between the *origin of life* and *evolution*. Then, evolution can be presented as information which supplies biological science with a central and continuing theme; it sheds light on the past, helps to explain the present, and offers hope and guidance for the future. As one of the important theories, it should receive due recognition and be discussed rationally and critically, with the requirement that students become aware of it, and of the unifying role it plays in biological science. These elements are compatible with the leading approaches to biological study today, but there should be no attempt to coerce students or to set up situations that would require that they believe in or give support to this or any other theory.

RAISING THE BIOETHICAL CONSCIOUSNESS OF BIOLOGY STUDENTS

Raising the bioethical consciousness of students is an important aspect of attitude development. Many hospitals, research institutes, and industries have *ethics boards* to curb unethical practices in the name of science, medicine, or scientific biological and technological progress. Practices are always under scrutiny. In one survey made, nearly 300 bio-medical research institutions were contacted and asked to respond on matters concerning the practice of experimentation with human subjects. The questionnaires, basically, asked individuals if they would approve some hypothetical practices or if they would consider them unethical, and the replies given suggest that better controls are required.

It is important that we provide our students with opportunities to assess their bioethical values. For this purpose, some of the hypothetical experiments (see sample, Figure 7-6) described in the prepared questionnaire are within their ability to understand and make a value judgment. While there are no absolutes, students can be encouraged to weigh the benefits that might accrue against the ethical invasion of the individual's rights in using humans for experimental purposes and make some moral judgment, always being aware of the two-edged nature of such practice, whether with or without patient consent.

In this respect, the use of animals as experimental subjects should also be considered, and visits made to pharmacological labs in research institutes where actual experiments are being conducted. Viewing rats in laboratory cages and comparing their care and treatment with those for which rat traps and rat poison are set to combat vermin, stem the spread of filth and disease, and control destruction to property should be assessed. Even the use of laboratory specimens in their own experiments should be scrutinized by students; such questions as "Was it necessary to raise fruitflies for later etherization in genetic studies?" and "Would you have learned as much from viewing a film devoted to the study?" should be raised and evaluated objectively.

A critical look at the Lifeboat Ethics question posed by Garrett Hardin* as a justifiable solution to the population problem facing man should be debated, as should the ethics involved in the pricing of prescription drugs, with considerations of chemical ingredients, costs of development, manufacture, marketing, and the incentive for future research and development. These are but a few of the areas that can serve as reminders to students that they should remain alert and responsive and that they should develop a consciousness of significant bioethical issues, whatever the encounter.

---

*Garrett Hardin, "Living on a Lifeboat," *BioScience* 10:561, October 1974.

A researcher plans to study bone metabolism in children suffering from a serious bone disease. He intends to determine the degree of appropriation of calcium into the bone by using radioactive calcium. In order to make an adequate comparison, he intends to use some healthy children as controls, and he plans to obtain the consent of the parents of both groups of children after explaining to them the nature and purposes of the investigation and the short and long-term risks to their children. Evidence from animals and earlier studies in humans indicates that the size of the radioactive dose to be administered here would only very slightly (say, by 5-10 chances in a million) increase the probability of the subjects involved contracting leukemia or experiencing other problems in the long run. While there are no definitive data as yet on the incidence of leukemia in children, a number of doctors and statistical sources indicate that the rate is about 250/million in persons under 18 years of age. Assume for the purpose of this question that the incidence of the bone disease being discussed is about the same as that for leukemia in children under 18 yrs of age. The investigation, if successful, would add greatly to medical knowledge regarding this particular bone disease, but the administration of the radioactive calcium would not be of immediate therapeutic benefit for either group of children. The results of the investigation may, however, eventually benefit the group of children suffering from the bone disease. Please assume for the purpose of this question that there is no other method that would produce the data the researcher desires. The researcher is known to be highly competent in this area.

Hypothetically assuming that you constitute an institutional review "committee of one," and that the proposed investigation has never been done before, please check the **lowest** probability that **you** would consider acceptable for **your** approval of the proposed investigation. (Check only **one**).

☐ 1. If the chances are 1 in 10 that the investigation will lead to an important medical discovery.

☐ 2. If the chances are 3 in 10 that the investigation will lead to an important medical discovery.

☐ 3. If the chances are 5 in 10 that the investigation will lead to an important medical discovery.

☐ 4. If the chances are 7 in 10 that the investigation will lead to an important medical discovery.

☐ 5. If the chances are 9 in 10 that the investigation will lead to an important medical discovery.

☐ 6. Place a check here if you feel that, as the proposal stands, the researcher should not attempt the investigation, no matter what the probability that an important medical discovery will result. **(If you checked here,** please explain): _____

Fig. 7-6 Hypothetical Experiment described in "The Ethics of Experimentation with Human Subjects" by Bernard Barber reprinted from SCIENTIFIC AMERICAN, Volume 234, No. 2, February 1976; with permission of the author

Providing students with opportunities to develop an awareness of the social, moral, and ethical implications of biological research and practices will help them to develop responsible attitudes toward the new techniques that will undoubtedly be introduced during their adult lives. By placing a high priority on the aspect of biological education that emphasizes responsibility for careful consideration of the underlying philosophy and basic concepts as well as the possible dangers and benefits that may be associated with technological advances, they will become better equipped to deal with techniques of the future that will probably, by today's standards, be more fantastic, more revolutionary, and more shocking than any that have been proposed to date.

8 Meaningful Use of Instruments and Techniques

Mere description and empiricism are no longer adequate for the teaching and learning of high school biology. Modern approaches to the study are more sophisticated and have come about primarily as a result of technological progress. Via applications of instruments and techniques developed in areas of physics and chemistry, biology has undergone an amazing evolution and has emerged as an experimental science utilizing the full spectrum of scientific processes and emphasizing problem-solving and open-ended investigations. Many achievements in biological science have been based on the application of precise laboratory techniques and their refinement for specific studies.

Bioinstrumentation is already well-established as a tool for analyzing the chemical composition of blood, studying brain waves, diagnosing certain diseases, and monitoring and analyzing heartbeat and respiratory rates and patterns. Biochemical techniques based on quantitative measurements are currently being employed for determination of food, air, and water quality; and there is reason to believe that the most significant changes that will affect life in the future will continue to be in the field of applied biology. The nature of the advances in biological science make it apparent that the employment of precise measuring techniques is essential for assisting the biologist in refining, extending, and supplementing his knowledge about life and its functioning on basic levels.

Instruments and techniques also contribute importantly to student investigations. They provide him with the means for collecting and analyzing quantitative biological data which might influence the outcome of his experiments and which will give to his study an authority that cannot be commanded by value estimations alone. But there should also be opportunities for him to view these technologies as something more than black boxes. Motivated by natural curiosity, the student should be further encouraged to open and explore the "black boxes" and come away from the experience with an understanding of what was being measured, how the measurement was determined, and how it relates to his investigative study.

All students can profit from exposure to basic demonstrations and investigations in which instrumentation and techniques are employed. Each tends to gain an appreciation of the practical value of certain technologies in revealing information about living systems that would not otherwise be possible for scientists to learn. The development of the student's concept that further scientific progress can be made through the use of known technologies also becomes evident. Further, the talented, science-oriented, and highly-motivated student can use basic knowledge and application of instruments for extending the open-ended investigations by way of in-depth studies and independent research projects in which the technologies can now be employed with skill and understanding.

BIOINSTRUMENTATION

Although some instruments have long been familiar in the biology laboratory, their number and usage is increasing.

Measuring with the Microscope

Size is an important characteristic of all living material, be it a cell, an organism, or any of their structures. Measuring that material may require only a rough estimate or a very precise measurement in very large or in very small units. For very small objects, the microscope becomes not only a device for viewing, but one for measuring as well.

Students can learn to make accurate measurements with the microscope, using an ocular micrometer that has been calibrated for a given microscope objective. To calibrate the ocular micrometer: carefully insert the ocular micrometer into the eyepiece of the microscope so that the engraved ruler on its surface can be read; place a stage micrometer having scribed lines that are precisely 0.01 mm. (10 microns) apart on the stage of the microscope and bring it into focus so that one of the scales is superimposed upon the other; line up both scales so that their first lines coincide, and locate an additional pair of lines that also coincide; count the number of ocular micrometer divisions and the number of stage micrometer divisions that are subtended by the two pairs; and calculate the magnitude of one ocular division. In Figure 8-1,

MEANINGFUL USE OF INSTRUMENTS AND TECHNIQUES   163

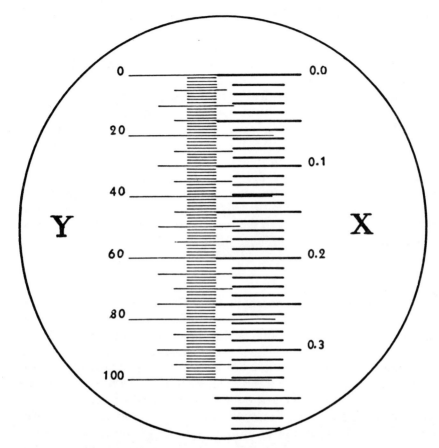

Fig. 8-1 Aligning the stage and ocular micrometers (Courtesy, American Optical Corporation)

for example, to calibrate the low power objective lens in which 7 stage micrometer (X) divisions line up with 21 ocular micrometer (Y) divisions, each ocular division = $7/21 \times 0.01$ mm. = 0.003 mm. or 3 microns. Students can use this reference to determine the length of various protozoans viewed, and to discover wide differences in cell size generally referred to as "microscopic."

In a similar manner, the high dry and oil immersion objective lenses can be calibrated to determine the magnitude of each ocular micrometer division when viewing with those magnifications.

After once having calibrated the ocular micrometer for each objective lens of his microscope, a student need never repeat the operation. He will have standardized his measuring device and will be able to employ it for making accurate microscopic measurements. From initial basic studies of determination of cell size, he may use the technique for collecting and analyzing quantitative data in experimental work involving the extent of skeletal muscle contraction after application of ATP (adenosine triphosphate) with and without ions, the elongation of cells from plant stems treated with various kinds and concentrations of auxins, and for assorted microbiological studies in which change in size is a critical factor.

Measuring the depth or thickness of a specimen under the microscope is accomplished with the aid of the calibrated fine adjustment screw. The ruled scale printed on its barrel indicates the magnitude of each unit. Thus, if a unit is indicated to be 0.02 mm., the distance turned in focusing from the upper to the lower surface of a cell being viewed will be its thickness, measured in units of 0.02 mm.

Use of the Spectrophotometer

The number of cells present in a yeast, bacterial, or other microorganismic culture is of concern to investigators pursuing studies involving the effects of environmental factors on growth rates and population density. During a study, it is frequently necessary to make a quick and easy determination of that number for an analysis of the study as it progresses, but the cost of a standard plate count is prohibitive in terms of both time and materials, particularly in cases where large numbers of counts or large numbers of cultures are involved. Fortunately, the turbidity of a broth culture gives an indication of the relative number of microorganisms present, and changes in its optical density are easily detected by a decrease in its ability to transmit light.

A quantitative measurement of turbidity can be made if (1) the intensity and wavelength of light entering the sample and (2) the amount of light passing through the sample (as recorded by a sensitive photoelectric cell) are known. Consequently, the spectrophotometer, used for precise measurement of optical density, is designed so that light from a light source passes through a sample to be tested before reaching the light-sensitive photomultiplier tube that converts the light into an electric current. The current, then amplified, is fed into a read-out meter that is calibrated in O.D. (optical density) units and corresponding percent of transmittance of light through the sample. The stronger the light reaching the photomultiplier tube, the greater the current produced by the tube and the greater the effect on the read-out meter. Since the more concentrated bacterial cultures allow less light to pass, the meter readings are in direct proportion to the density of the culture being tested, and O.D. values and percent of light transmittance can be read at one and the same time on the meter scale.

The spectrophotometer can be demonstrated by zeroing the instrument with a blank (uninoculated broth media only) placed in the tube for light to pass through. The meter should read O.D. = 0, 100% light transmittance. To demonstrate: prepare a series of dilutions of 24-hour broth culture of *Escherichia coli* in some series (example: undiluted, 10^{-1}, 10^{-2}, 10^{-3} . . .); zero the spectrophotometer using a blank; following the protocol for the instrument model and recommended wavelength, determine the O.D. and/or percent of light transmittance for each sample; correlate the O.D. and/or percent of transmittance of light with the visual turbidity and expectations for the sliding scale dilutions; then calibrate the readings by sampling each of the dilutions for a conventional pour plate-count technique determination of the number of organisms/ml. of each dilution. The actual number of organisms/ml. should be recorded for each O.D. value obtained from readings. Once calibrated for a given wave length, broth, and organism, the O.D. readings are permanent, reliable representations of the population densities, and a more economical method for making these determinations in a study that requires population density readings for numerous samples. Using this technique, students come to understand the value of the instrument, the principle of its operation, and the ways in which it can be employed in the determination of effects of environmental factors, age, and nutritional components that may be encountered as parts of an independent investigative study.

Lung Capacity and the Spirometer

Measurements of lung capacity are good indicators of an individual's respiratory health; a lowered volume of air exchange with the atmosphere may indicate diseased tissues which decrease the ability of the lungs to fully expand or a weakening of the muscles associated with the respiratory mechanism.

Students may demonstrate, by differential expiration, the Tidal, Supplemental, and Complemental air volumes which comprise the Vital Capacity, or total air volume that one is capable of expelling from one's lungs. This is the most convenient, and therefore the most commonly measured air volume for comparative lung capacity studies.

A fairly good estimate of Vital Capacity (or V.C.) can be obtained using a large jar marked in graduated 100 ml. divisions, such that when filled with water and fitted with one end of a length of plastic or rubber tubing inserted into its open end, this part of the assembly can be inverted and completely submerged in a tank of water, with the other end of the tubing remaining free. A subject, expelling from his lungs *all* of the air of which he is capable (a residual air volume of about 1 to 1½ liters remains in the lungs to prevent their collapse) with one complete expiration, exhales this air into the free end of the tubing equipped with a mouthpiece, and can observe the volume of water in the graduated jar which it displaces. Hence, the Vital Capacity, generally ranging from 2800 to 5800 ml. among high school students, can be measured, and the rationale of the method employed in its measurement established.

For a more accurate measurement of Vital Capacity, a Spirometer, with V.C. volumes that can be read on a metered scale, is preferable. Individuals tested should also calculate their Vital Capacity based on considerations of age, sex, and height, using the formulas

Male V.C. = {[.052H(cm.) − .022A(yrs)] − 3.50} × 1000 = V.C. in ml.
Female V.C. = {[.041H(cm.) − .018A(yrs)] − 2.69} × 1000 = V.C. in ml.

On the basis of the data collected, it should be noted that it is the relationship existing between the calculated and the actual Vital Capacity that is significant. Thus, if a boy calculates his Vital Capacity to be 5000 ml. but registers only 4800 ml. with an actual Spirometer measurement, he can compute the relationship

$$\frac{\text{Actual V.C.}}{\text{Calculated V.C.}} = \frac{4800}{5000} = .96 \text{ or } 96\% \text{ of lung capacity being utilized}$$

These percentages can then be researched for interpretation.

Students are highly motivated by such a demonstration and by the investigative studies which it suggests. Comparative studies of athletes vs. non-athletes, smokers vs. non-smokers, girls vs. boys, and athletes engaged in running sports vs. those engaged in body contact sports are among the most popular, with students free to design and conduct others.

Demonstrating the Sphygmomanometer and the Biomonitor

Records reveal that an alarming 54% of the total number of deaths in the United States in 1974 were attributed to diseases and disorders of the heart and blood vessels. This implies a need for all to acquire knowledge and understanding of the effects of various factors on patterns of normal functioning. For the purpose of collecting pertinent quantitative data, some familiar instruments are commonly employed.

A sphygmomanometer is used to measure in millimeters of mercury the pressure that blood exerts on the walls of the arteries—greater when the heart muscle contracts (a systole) than when it relaxes (a diastole) between beats. Hence, blood pressure really involves two separate readings: Systolic Pressure/Diastolic Pressure, for example, 120/80.

The method of measuring blood pressure employs the sleeve of a sphygmomanometer for applying just enough pressure to the brachial artery so as to stop the flow of blood through it. At this point no sound of blood flow can be heard through the stethoscope when held over the artery, just below the inflated sleeve. If the air is slowly released from the sleeve, the pressure on the arterial walls caused by the wave of blood that has just been ejected during the pumping of blood by the ventricle can be detected by the sounds heard through the stethoscope and by pressure readings recorded on the dial of the sphygmomanometer. This is the *systolic* pressure. As more air is allowed to escape, the sound of the blood pressing on the arterial wall disappears and identifies the pressure exerted on the walls as the heart relaxes, allowing the ventricles to fill up with blood for the next cycle. This pressure is called *diastolic*.

Blood pressure under normal conditions should be demonstrated and studies planned to investigate the effects of various stress conditions (exercise, standing for a prolonged period of time, anxiety, etc.). Measurement of blood pressure under these and other conditions helps students to gain some insight into the causes—and dangers—of hypertension, strokes, heart attacks, atherosclerosis, cholesterol build-up, conditions of overweight, and other medical conditions, and may be conducted by student groups, if properly supervised and monitored.

There are also many factors that affect the rate at which the heart beats; for these studies, the Biomonitor* is a useful device. This electronic instrument, basically a millivolt amplifier, is designed to detect and display the bioelectric events associated with cardiac muscle activity. After becoming adept with the operation of the instrument and determination of the mean heartbeat of a subject, students can extend the use of the Biomonitor for gathering useful quantitative data in investigative studies of heart rate differences as determined by height and sex, and of the effects of holding the breath, of rapid breathing, and of exercise on heart rate.

The Kymograph for Physiological Studies

The Kymograph is a useful and versatile tool of the physiologist. Not only does it give accurate and precise measurements of the rates at

---

*An instrument for physiological studies by Phipps & Bird, Inc., Richmond, Virginia.

which physiological processes are occurring, but it also provides a record in the form of a graph. The Kymograph is easy to operate; it consists of an electric-driven motor that powers a revolving drum mounted on a vertical post, and the speed can be controlled to a specific number of revolutions per minute. The drum is covered with a removable graph paper on which a stylus records lever activity caused by impulses from the attached sensor in contact with the experimental subject. Once introduced to the operation of the Kymograph, students have an impressive and professional tool for engaging in physiological studies.

With a Pneumograph attached to the chest area of a willing subject, respiratory rates can be determined for his normal breathing while at rest and for modified respiration after exercise, while holding the breath for 60 seconds, while breathing into a paper bag for 60 seconds, and while laughing, sneezing, coughing, or drinking a glass of water. Respiratory activity is reflected in chest movements which cause impulses that are recorded on the moving drum as tracings. When the activity has been completed, the tracings under various conditions can be used for an analysis of each factor and its influence on the normal pattern.

Using the Ergograph, the physiology of muscle fatigue can be studied. With this attachment, a pistol grip action is provided with a sliding fulcrum control for the regulation of stroke amplitude and, by working the index finger against an adjustable coil spring, fatigue curve tracings on the graph can be analyzed and followed up with a research study of the biochemistry of muscle fatigue.

Students may engage in other studies suggested by the use of the Kymograph. Possibilities include the physiology of frog muscle contraction and experiments involving the effects of temperature and/or chemical substances on the heart rate of frog or turtle where this first-hand collection of data is needed for individual investigative work.

BIOCHEMICAL TECHNIQUES

The biologist cannot fully understand the functioning of living things without some knowledge of chemistry. However, once revealed, the biochemical processes are so well-organized as to make the organisms engaging in them predictable performers and, as such, useful for bio-assays. In the rapidly expanding field of biochemistry, there are applications to many practical situations.

Determination of Dissolved Oxygen in Water

Water pollution is basically a biological problem. It is most noticeable in the destruction of aquatic animal life where, to a large extent, the destruction is due to a deficiency in the dissolved oxygen content of the water. An analysis of water pollution, whether employing a direct measurement of dissolved oxygen by the Winkler method or by the indirect BOD (biochemical oxygen demand) test, is essentially concerned with a quantitative determination of the dissolved oxygen (DO) in the water. A modification of the Winkler method makes it possible to collect and "fix" the sample at its natural site and test it in the laboratory.

In the *field*, collect the sample and prepare it for titration:

1. Fill a 250 ml. glass-stoppered reagent bottle with the water sample, allowing no exposure to air or trapped bubbles in the bottle.
2. Add 2 ml. of manganese sulfate solution (48 g. $MnSO_4$ in 95 ml. boiled water, filtered and diluted to volume of 100 ml. in distilled water) with pipette inserted below water surface level.
3. Similarly add 2 ml. alkali-iodide-azide solution (15 g. KI and 50 g. NaOH dissolved in boiled distilled water and diluted to a volume of 100 ml.; add 1 g. NaN_3 dissolved in 4 ml. boiled distilled water).
4. Stopper sample bottle and invert several times to mix. Allow no air bubbles. Allow precipitate to form and settle. Then shake again and allow to settle.

In the *lab*:

1. Add 2 ml. concentrated H_2SO_4 to sample bottle and invert several times to dissolve precipitate.
2. Pour contents of bottle into 500 ml. Erlenmeyer flask. With buret, slowly titrate 0.0375M sodium thiosulfate. (Dissolve 9.3 g. $Na_2S_2O_3 \cdot 5H_2O$ in freshly boiled, then cooled, distilled water and dilute to volume of 1 liter.)
3. Add 2 ml. starch solution. (Add small amount of distilled water to 5 g. soluble starch to make paste. Add paste to 1 liter of boiling distilled water and continue to boil for 2 to 3 minutes. Cool, cover, and allow to settle. Use clear supernatant.) The appearance of a blue color indicates the presence of molecular

iodine. Continue titrating, while agitating the sample, until the blue color disappears. This indicates that molecular iodine has been reduced to iodine ions.
4. Record the total volume of titrant used and calculate the DO (dissolved oxygen) in ppm.
   Example: Each ml. of 0.0375M sodium thiosulfate used in the titration of a 250 ml. water sample indicates the presence of 0.25 mg. O_2 per 250 ml. sample. This may be expressed as 1 mg. O_2 per liter, or 1 ppm (part per million).

Interpretation of results in terms of acceptable levels for the area should be researched. Also, samples representing different water levels, different locations along a stream or waterway, different temperature levels, and samples taken from the same location at different times of day or seasons of the year make interesting studies for ecology-minded students. If studies are undertaken, they can be demonstrated to ecology classes, and, if local areas are investigated, the interest level of both demonstrator and audience is unusually high.

Using the Thunberg Technique

During the oxidation of an organic substrate, hydrogen atoms are removed from the substrate and are accepted by another substance which then becomes reduced. The enzyme dehydrogenase plays an important role in this biochemical process, and in respiration the hydrogen is transferred to gaseous oxygen by several pathways. However, dehydrogenation will take place in the absence of oxygen if a suitable hydrogen acceptor such as methylene blue is present. In this case, the blue dye is reduced to a colorless leuco compound.

$$AH_2 + MB \longrightarrow A + MBH_2$$

AH_2	+	MB	⟶	A	+	MBH_2
substrate		methylene blue		oxidized substrate		colorless leuco compound

The rapidity of decolorization of the synthetic dye is an indication of the rate at which oxidation takes place. If bacteria are the test organisms, this rate is an indicator of their dehydrogenase activity, and it forms the basis for the Thunberg technique: methylene blue is added to a bacterial suspension in an evacuated test tube, and the dye is allowed to slowly become decolorized due to the oxidation of cell reserve food materials; when the bacteria are provided with a suitable oxidizable substrate, the rate of color disappearance with the different substrates used can be measured as an indicator of the relative rates at which the substrates are metabolized by the test organisms.

Students can perform a variation of this technique which estimates the number of bacteria present in milk samples; 1 ml. of methylene blue (1:20,000) is added to each of a series of 10 ml. milk samples of various ages, sources, or other conditions. Tubes containing well-mixed milk and methylene blue are stoppered, placed in a 37°C water bath, and examined at intervals for signs of decolorization indicating dehydrogenase activity of microorganisms present. The rate of decolorization is determined by the amount of time required for at least $4/5$ of the tube content to turn white and is related to the number of microorganisms present to produce the biochemical change. Classification of milk on the basis of the number of bacteria per ml. necessary for the reduction of methylene blue in specified time periods should be used for reference and for interpreting the data collected.

The determination of the number of bacteria present in a unit quantity of milk uses an indirect approach. Highly-motivated students should be encouraged to make a determination by direct means and to compare the results obtained by the two methods. Having established the reliability of the Methylene Blue Reductase Test, they can then apply the technique to the determination of number of bacteria per ml. in bacterial broth cultures grown under varying conditions of light, temperature, mineral and vitamin supplements, and with various concentrations of additives suspected of having species toxicity.

MULTI-TECHNOLOGICAL APPROACH

The dynamics of living systems is such that their study requires the employment of many different procedures; instrumentation, mathematical calculation, and biochemistry may be used in combination when seeking knowledge and understanding of life and the activities associated with it.

Investigating Metabolism

The impact of some environmental factors on the physiological adjustments which an organism must make to maintain its homeostasis can be measured in the metabolic rates of small animals. The concept is based on the fact that all organisms require organic molecules for their oxidative processes in which heat is produced. The amount of oxidation depends upon the cells and tissues of the organism so that a measure of the heat produced per unit area of tissue per unit of time while the organism is at rest constitutes the *basal metabolism*. Put another way, the basal metabolism may be described as the level of metabolic activity required for the maintenance of the automatic functions of the body.

Metabolic rate refers to the metabolism in a given period of time, for example, one hour. It can be estimated from (1) food consumption, (2) energy released in the form of heat, or (3) the amount of oxygen used in the oxidation processes employed to yield energy. Generally, the smaller the animal the higher its metabolic rate—and a high metabolic rate requires a correspondingly high food intake and rapid digestion, circulation, respiration, and heartbeat. Hummingbirds and shrews exhibit this phenomenon; the shrew, for example, weighs about 4 grams and must engage in a constant search for food to accommodate its excessively high metabolic rate. During periods of food deprivation its body reserves are rapidly depleted and, if relief does not come quickly, it will starve to death. Developing this concept, students can see how the shrew reaches the ultimate in smallness for a mammal.

An accurate measurement of the respiratory rate of an organism is a first step in the serious study of physiology because the respiratory rate is one of the most recognizable outward signs of its metabolism. If we can measure respiratory rates, we can also determine what factors affect metabolism. The influence of temperature, drugs, poisons, and caffeine-containing beverages on metabolism can be investigated by means of a Respirometer (see Figure 8-2).

There are many designs for respirometers, as befits the needs for accommodating the microorganism, plant, poikilothermic animal, or mammal to be tested. Respirometers for small animals generally consist of a metabolic chamber fitted with an animal cage, area for soda lime, large rubber stopper fitted with a calibrated glass tube, and a thermometer for use in the chamber. After a layer of soda lime has been introduced to the floor of the chamber (to absorb all expired CO_2), the test animal should be weighed, placed in the cage, and allowed to rest quietly for 10 to 15 minutes while temperature equilibrium is reached in the chamber. Then the inside of the calibrated tube should be wetted with water and, after the tube and stopper are secure and in proper position to close the open end of the chamber, sealed with a drop of detergent or soap bubble solution placed at the open end of the tube. Using a stop watch to determine the time in seconds required for the bubble to travel along a measurable distance (5 ml. calibration on a pipette), repeat the procedure and record the time for several runs. A simple calculation will yield a figure indicating the milliliters of oxygen utilized per minute.

Since the production of heat by an animal is related to its total surface area, we calculate the square measure of the animal by covering its body with foil and measuring its area in square centimeters or by

MEANINGFUL USE OF INSTRUMENTS AND TECHNIQUES 173

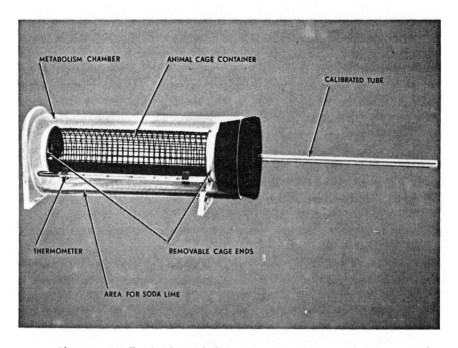

Fig. 8-2 Small animal metabolism apparatus; courtesy of Phipps & Bird, Inc., Richmond, Virginia 23261

reference to available listings* of standard surface area/weight relationships for small animals. If we then assume that a normal small animal will release 4.8 kilocalories of heat for each liter of oxygen (or 4.8 gram calories of heat for each milliliter of oxygen) used, the gram calories per hour can be corrected for standard conditions of temperature and pressure and computed:

$$\frac{\text{Av. ml. } O_2/\text{min.} \times 4.8 \text{ gram cal/ml. } O_2 \times 60 \text{ min.}}{\text{Surface area in cm.}^2} = \text{gram cal/hr/cm}^2$$

Example: O_2 used/min. (corrected) = 3.2 ml.
gram calories/ml. O_2 = 4.8
surface area = 75.4 cm^2
60 min. = 1 hr
calculation: $\frac{3.2 \times 4.8 \times 60}{75.4}$ = 12.2 gram cal/hr/cm^2

The heat produced to maintain metabolic needs will vary and can be studied under conditions following food intake, exercise, excitement,

---

*Chart for Rat or Gerbil in *Small Animal Metabolism Apparatus Instruction Manual*, Phipps & Bird, Richmond, Virginia.

or some other reasonable experimental factor. Students' should, of course, be advised of proper care and handling of all small animals used in laboratory work.

Detecting Water Pollution

The Sterifil* technique is used worldwide for performing tests on water samples. With it, water is tested for the presence of coliform bacteria because they are almost always found in raw sewage where disease producing bacteria may also be present. Hence, the coliforms, because of their easy identification and because of the company they keep, are used as indicators—not causers—of water pollution.

The microbiological techniques involved in the testing are easily handled by students. They must, of course, maintain aseptic conditions and practice the outlined protocol:

1. Sterilize all components of the STERIFIL apparatus.
2. Dilute the sample to be tested, using sterile distilled water, and pass the sample through the apparatus employing a 0.45 μ millipore gridded filter. Organisms larger than pore size will be retained on the gridded surface.
3. Transfer filter to prepared MF Endo medium in a culture dish and incubate in an inverted position at 37°C for 24 hours.
4. Examine colonies grown on surface of filter membrane.
5. Count the number of green sheen colonies and calculate the number of organisms/ml. of original sample.
6. Check local standards for interpretation of results as they relate to the quality of the water sample.

The biochemistry of coliforms grown on MF Endo medium should be well-established as it relates to their identification: the medium contains, in addition to basic nutrients, lactose and a basic fuchsin stain (usually red, but here made pink by the addition of sodium sulfite). Most bacteria coming in contact with this stain will be affected by the dye and appear red. However, coliforms alone have the ability to degrade lactose into simpler substances, among which are some aldehydes whose molecules attach to molecules of the fuchsin-sulfite complex and produce a green sheen by which their colonies can be identified. The sequence of events is easy for students to follow: since only coliforms form aldehydes from lactose, only coliforms produce green sheen col-

---

*A complete sterile filtration system is available from the Millipore Corporation, Bedford, Massachusetts.

onies, and all other bacteria remain undifferentiated in a mixed culture.

Sterile filtration and selective media for growth of microorganisms can be used in combination for other applications of the technique; determination of the presence of bacteria in various air, fruit juice, and milk samples, and effects of temperature and/or disinfectants on bacterial growth make challenging and interesting studies, while yielding quantitative data for analysis.

Tracing the Pathway of Carbon in Photosynthesis

It is extremely difficult to observe the physiological processes by which organisms maintain their integrity, and often it is impossible to duplicate them in an *in vitro* situation. This is well-established for both plant and animal systems in which the biochemistry involved is unique to the living substance and somewhat elusive to the investigator, yet vital to his understanding of the processes and the manner in which variable factors will affect them in both qualitative and quantitative ways.

Because nuclear changes do not affect the chemistry of an element, radioactive isotopes do not act significantly differently from other atoms of their kind except in their ability to send out signals indicating their presence. This is an advantage to the researcher who, using a device for detecting the signals, can then trace their pathways through an organism. The Geiger-Mueller tube is both a suitable and convenient device. It consists of a metal cup with an electrically-insulated wire running down its center and a thin mica window (a point of entry for high energy radiation particles) covering its open end. If, in a test run, an electrical potential of about 1000 volts is applied between the negatively charged cup and the positively charged central wire, any high energy beta particles entering the tube are drawn by the potential to the wire, creating a flow of current that is then fed into an appropriate electronic device and registered on the dial. Correcting for background radiation, the radioactive substances can be located and traced in a physiological process *in vivo*.

Photosynthetic fixation of CO_2 can be demonstrated with the use of $C^{14}O_2$ and variegated (green-white) *Coleus blumei* leaves in a closed chamber which allows them to be exposed to light. To demonstrate: Place 10 microcuries of radioactive barium carbonate in a 500 ml. Erlenmeyer flask and carefully add several drops of 1M H_2SO_4, making sure that the acid drops directly to the bottom of the flask to initiate the generation of $C^{14}O_2$:

$$BaC^{14}O_3 + H_2SO_4 \longrightarrow BaSO_4 + H_2O + C^{14}O_2$$

Ring the top of the flask with vaseline and press a green-white section of a Coleus leaf over the flask opening, securing it to the vaseline along the rim to effect an air-tight seal; allow the leaf to remain in position for 20 to 30 minutes, then gently remove the leaf and, with a cork borer, punch out circular discs from both green and white areas; transfer the punched out discs to individual petri dishes and, employing a Geiger-Mueller tube, record the number of counts for each sample over a 10-minute period. Correct for background radiation, and determine the rate of carbon dioxide fixation under conditions of presence and absence of chlorophyll.

C^{14} has assisted investigators to determine the exact pathways of the carbon in the CO_2 molecule and to learn how rapidly it is used under various conditions. Using these techniques, and properly supervised while using radioactive materials, students can pursue investigations of CO_2 fixation under various conditions of light intensity and/or temperature, or they may research the pathways of other elements using radioactive isotopes such as P_{32} in plants and animal nutrients. The technique can be extended to include autoradiography as an alternative tracing method, but care should always be taken to practice recommended handling procedures during the activity.

The employment of some instruments and techniques in a biological study may not be planned. Students engaging in an independent project may have need for quantitative data collection and/or analysis that might reasonably involve:

- Chromatographic or electrophoretic analysis of leaf pigments.
- Potometer-determined rate of transpiration by plants.
- Direct microorganism or blood cell counts with a counting chamber.
- Analysis of bioelectric wave patterns revealed by use of an oscilloscope or IMPScope.*
- Analysis of effects of photoperiods on plants grown in an environmental chamber.
- Interpretation of the hematocrit measure of the formed elements in blood.
- Analysis of results of biochemical testing involving the effects of various concentrations of antibiotics on a sensitive organism.

---

*This combined amplifier, oscilloscope, and tissue stimulator was developed in an NSF project devoted to Instrumentation Methods for Physiological Studies.

- Computer identification of microorganisms based on basic morphological parameters.
- Statistical analysis of a *Drosophila* trihybrid cross using a computer program to simulate the breeding pattern and produce the desired large number of offspring.

Involving students in investigations that place emphasis on the quantitative aspects of biological study can enhance their awareness of the trend toward replacing the human element in making observations with the more objective and accurate measurements obtained by instrumentation. Through experiences which highlight measuring things with the precision and reliability afforded by tools and techniques of the same kind developed by and for chemists and physicists, students come to develop an understanding and appreciation of the importance of some sophisticated methods used in many areas of experimentation. It is then that they begin to perceive biology as a science that goes beyond the mere descriptive phase to one in which progress in technology contributes to further progress and understanding.

9 Evaluating Student Progress

Concomitant with the increased student involvement in the planning and development of their studies in biological science, students are also becoming more acutely aware of and concerned with how well they, as individuals and as a group, are able to demonstrate their ability to understand and use information about topics they are investigating. As is now recognized, performance on content-oriented tests alone does not provide an adequate measure of a student's knowledge and ability to use information in a meaningful way; the implication, then, is for the development of dynamic and effective techniques for evaluating the kind and amount of teaching and learning that is taking place and for bringing about improvement in both.

In developing a workable program, we should view evaluation as an integral part of the teaching-learning process. It should provide students with opportunities for an on-going assessment of their progress in the course, facilitating their efforts to summarize and interpret what they are learning, to identify specific areas in need of improvement, and to engage in a continual resetting of goals via recognition of readiness to replace those achieved with more challenging ones at a higher level. It should also equip us with powerful tools for assisting and guiding this development and for identifying individual levels of achievement for which grades must be assigned.

DESIGNING AN EVALUATION PROGRAM

To ensure that the means by which learning and achievement is measured will be varied, interesting, and optimum for each learning activity, we must develop and employ strategies which are dynamic in approach and diverse in design. When used in combination, they should properly accommodate every aspect of a student's learning and assist in the development of his capacity for self-evaluation, while enabling us to make a valid appraisal of his individual achievement. The evaluation program must certainly involve more than a routine administration of tests for grading purposes. It is important that we distinguish the various "evaluations" and, for each, employ the most effective techniques:

- Contractual fulfillments can be used to identify the level of accomplishment attained by a student in accordance with the degree of difficulty of specific tasks selected by him for his "contract" agreement.
- Individual reports, both oral and written, provide opportunities for a student to indicate his depth of understanding of a topic and his ability to express and convey the knowledge to others.
- Individual lab reports provide opportunities for students to demonstrate their ability to collect and analyze data, interpret results, draw conclusions, and make practical applications of their findings in a hands-on situation. They provide an identification of goals achieved and serve as a springboard for engaging in extensions of completed work, as indicated by readiness for resetting goals in an open-ended investigative study.
- Daily quizzes at the beginning or end of a class session enable students to succeed in small segments and give them encouragement for putting it all together. With immediate feedback, the technique reinforces learning and permits remedial work for correction of errors before larger units of work over longer periods of time have compounded many small errors into gross misunderstandings.
- Small group interactions permit students freedom to discuss a topic under study, with opportunities for each to contribute to the discussion, during which he gains insight into his personal level of understanding, to assess his individual attitudes toward social, moral, and ethical issues that may be involved, and to evaluate his own accomplishment, with perspective.
- Alternate forms of tests provide students with opportunities to pinpoint areas of weakness on the first one administered, for remediation prior to experiencing the second form for rechecking.
- The use of pretests allows for identification of previous knowl-

edge for use in planning and directing the course of study for a class or an individual.
- Dialogues, one-to-one encounters, and discussions with small groups permit a personal exchange in which students have opportunities to exhibit individual levels of comprehension. Because the technique allows the teacher to probe more deeply as responses are made, it is a useful device for establishing the upper limit of the student's ability to understand the principles and concepts involved in a study under consideration.
- Open book and/or notebook tests assess student ability to find and use information in a meaningful way. Questions for thought and expression and problem solving situations based on factual information test his comprehension and ability to make effective use of reference materials.
- Teacher-made tests and quizzes allow students to exhibit, via specially designed evaluation forms, the extent to which goals and objectives have been achieved for a cooperatively developed learning activity.
- Standardized tests can be used to determine the relative placement or percentile rank of a student as compared with established norms for performance on a test administered on a wide scale.

VALUES OF OPPOSING EVALUATION FORMS

There have been many controversies concerning the relative merits of factual vs. interpretative evaluations, short form vs. full expression essay answers, and oral vs. written test forms. Upon examination, we find that each has value and, when applied appropriately to the purpose it is designed to serve, can contribute importantly to the overall program of evaluation.

Guidelines for Use of Opposing Evaluation Forms

FACTUAL	vs.	INTERPRETATIVE

When significant test factor is:
  mastery of vocabulary
  knowledge of biological terms
  identification of concepts and principles
		When test factor involves use of information in:
		  reading graphs
		  interpreting data collected
		  analyzing diagrams

When purpose of test is for evaluating retention of learning

When immediate feedback for reinforcement of learning is a key factor

When test purpose is for evaluation of comprehension as evidenced by: ability to apply information to the solution of a problem or to a new situation; ability to make predictions based on available information

SHORT ANSWER vs. ESSAY-TYPE ANSWER

When large number of test items are to be included

When same test applies to a relatively large number of students

When time required for test construction is not a limiting factor

When time for grading papers is a limiting factor

When items being tested lend themselves to multiple choice, matching, or other precise answer form

When immediate feedback to students is a critical factor

When relatively few questions are to be included

When topics may be highly specialized or individualized

When evaluation form is designed to allow students freedom to organize and develop an answer according to their own style.

When personal interpretation and subjectivity in answers are desirable

When time required for test construction is limited

When time required for reading and grading papers is not limited

When it is not crucial that students have papers returned immediately

ORAL vs. WRITTEN

When used for spontaneous evaluations

When interacting with individuals or small groups to determine readiness to proceed to next level

When summarizing or reviewing for a major topic test

When it is important for students to gain familiarity with

When it is desirable for students to have copy of evaluation sheet for individual progress or notebook record

When it is important that every student respond to every question posed

When the same evaluation is to be used for absentee student make-ups

pronunciation and verbalization
of terms and expressions

When students share findings
about their investigations with
others in reports and discussions

When laboratory and research
findings are being reported and
conclusions are being evaluated

INNOVATIVE EVALUATION STRATEGIES THAT WORK

At a recent symposium on evaluation procedures used by high school biology teachers, personal experiences with some innovative practices and variations of some familiar strategies were reported to have merit:

Games, either commercially-prepared or teacher-made, are especially helpful for evaluating student knowledge of vocabulary and their ability to identify and associate biological structures, processes, and interrelationships. Students like playing games, and the incentive for improvement is great because the remediation is immediate and the re-learned knowledge is used at once—and repeatedly—for mastery. There have been many reports that, after having employed word games and puzzles as a testing device, either to replace the traditional matching exercise or as a bonus question with extra points scored for correct solutions, student vocabulary performance test scores were observed to show noticeable improvement.

Essay questions collected from previously administered Advanced Placement Tests are excellent sources of material for student topic papers. They serve as models and practice materials for helping students to grasp something of the depth of treatment and scientific support needed for imaginative situations as well as classical studies in biology. Additionally, they reveal the extent to which students can analyze, synthesize, and interrelate basic concepts involved in a given situation. On the basis of individual readiness, students can then apply careful thought and clear expression to course-related, teacher-constructed topics:

Example: "You are a biological engineer, commissioned to work on a project design for the year 2080. List the events in the course of a day of a 16-year-old student of the future and give reasons for his activities and the way they will be conducted. Keep in mind the technological changes that will probably have been made by that date and, in your projection, give attention to matters of food and energy, health habits, school and leisure time activities, travel, living accommodations, and the state of the environment."

False statements requiring that corrections be made in order to make them true are better evaluators than those calling for a true or false determination only. Italicizing or underlining the key word or phrase in each statement focuses student attention on the critical issues and forces the student to be more discerning in making responses.

Example: Some statements appearing below are true; others are false. Indicate a true statement by writing the word TRUE in the space to the left of the statement; if, however, the statement is false, make changes to make it true by substituting a word or group of words to replace the words in italics:

(fat bodies)   1. The frog stores body fat in *layers under the skin.*

  TRUE   2. The aorta in the frog is an artery that carries blood *away from* the heart.

―――――   3. . . .

Teachers using this variation of true-false questioning report that their satisfaction derives primarily from the fact that it discourages guessing. Further, it encourages students to read with a critical approach which, as an added bonus, they carry over to other readings in biology.

Interpretations of biological topics in terms of quotations from song lyrics, literature, and advertising slogans, as well as quotes by famous personages offer stimulating and challenging evaluations for students to ponder:

Examples: Explain how an alteration of only 2 nucleotides of the DNA molecule can lead to a cascade of consequences resulting in a genetically-determined condition according to the pattern expressed by George Herbert's ". . . for the want of a nail the shoe is lost, for the want of a shoe the horse is lost, and for the want of a horse the rider is lost."

Interpret the advent of Sir Alexander Fleming's "chance" discovery of penicillin in terms of the reference to Louis Pasteur's successes, often quoted: "In the fields of observation, chance favors only the prepared mind."

Comment on the flaw in the use of selected sperm alone to regulate breeding as pointed out by an incident involving the beautiful dancer Isadora Duncan, whose proposal of marriage to the brilliant playwright George Bernard Shaw,

EVALUATING STUDENT PROGRESS 185

"Think of our children—with my body and your brains," was met with Mr. Shaw's reply, "But what if they have my body and your brains?"

Activity type questions lend variety and a change-of-pace to testing procedures. Posing situations which test a student's ability to solve an original problem, make a practical application of basic information, or use information in a personal way are both stimulating and challenging—and the more practical and personal the situation, the greater the motivation. Students exposed to this form of evaluation learn to demonstrate their ability to apply theory to new situations which, for them, are meaningful.

Example: The ability to taste certain chemical substances is due to a dominant gene (T). Using the taste papers attached below, perform a taste test for the three chemical substances and determine your phenotype and genotype for ability to taste each.

Chemical Taste Substances	Test Papers Attached	Phenotype	Genotype
Phenylthio-carbamide	test paper		
Thiourea	test paper		
Sodium benzoate	test paper		

Questions based on graph analysis of data collected in an experimental study are more effective evaluators than are those based on mere word descriptions. They go beyond testing on the knowledge level and ask students to apply what they know and to interpret the data in a meaningful way. Data from actual laboratory investigations as well as hypothetical situations can be used in challenging ways.

Example: Using information gathered from first-hand experiences in a student-performed experiment, the following graph was constructed:

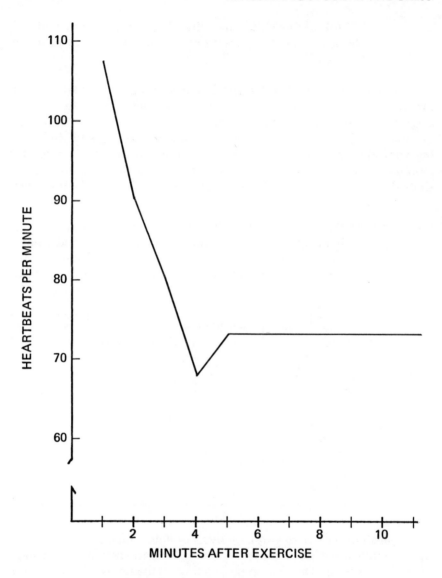

Fig. 9-1 Heartrate recovery after exercise

Analyze the graph and answer the following questions:

1. By how many beats/minute was the heart rate increased by exercise?
2. How many minutes were necessary for the normal heart rate to be restored?

3. What was the apparent normal heart rate for the individual tested?
4. During which minute was the greatest adjustment made to restore the heart rate to normal?

Presenting a problem for students to solve according to some principle developed during the class or lab session encourages them to think and to apply what they have just learned. Through the immediate employment of information from the lesson, the technique both stimulates and reinforces student learning.

Example: Following a lab session devoted to a study of members of the worm phyla—Platyhelminthes, Nemathelminthes, and Annelida—students may be presented with an "unknown" specimen, such as an *Aeolosoma hemprichi* for classification based on its structural characteristics. The technique can also be applied to other appropriate groupings, such as orders of insects, fresh water algal forms, or basic bacterial types.

35 mm. slides used to present questions on a laboratory evaluation ensure that all students will see the specimen, lab setup, or microscope view at its optimum, no matter when the viewing takes place. Slides are as applicable to laboratory measurements, categorizing and identifying specimens and models, interpreting experiments, and identifying appropriate interrelationships of phenomena and ideas as is the more traditional "20 stations" lab-practical test—and with some important advantages. The slides have value for reuse throughout the test day and for later use when retesting, for makeups, and for year-to-year use, where appropriate.

BLUEPRINT FOR CONSTRUCTING EXAMINATIONS AND UNIT TESTS

There are two basic reasons for developing and administering teacher-made examinations and tests for major units of work in biology: (1) they offer the best and most reliable evaluations of how well students have mastered certain information and achieved the course objectives, and (2) unless students are required to use the entire scope and range of content and approach that the course involves, they are apt to conclude that the course objectives are really not important.

Whether used at the completion of each major topic study or for integrating all aspects of the course at the end of the semester, the testing should be designed to measure each student's knowledge, and should highlight his ability to use, interpret, apply, and communicate its meaning. Further, it should provide a capsule account of the course in such a way as to enable the student to put it all together and to assess for himself the degree to which he experiences personal satisfaction with his accomplishment.

Constructing such a test is a demanding and time-consuming task. However, it can be facilitated by the use of an organized approach, following a step-wise procedure:

1. Determine the scope of the test by listing the major concept areas to be included.
2. Determine the content of the test by listing, within each concept area, all principles, understandings, interpretations, applications, and relationships deemed essential for students to know.
3. Determine, for each of the content items listed, the most effective way to evoke responses and answers from students.
4. Group items into categories, as indicated in (3), and write questions for each item in accordance with the corresponding protocol.
5. Arrange questions, within each category, in order of increasing level of difficulty and/or complexity.
6. Check the test, item by item, category by category, to ascertain that:

   - directions for each section are clearly stated
   - there are no ambiguities in wording or expression
   - variety and balance is achieved via types of question included
   - there are no "trick" questions or trivia included
   - the test is properly addressed to the course objectives
   - if there are duplications of items designed to serve the same purpose, students should be allowed to respond to one only, on a choice-selection basis
   - the test is fair and it is possible for all students to achieve a passing grade
   - the test offers a challenge to students of all ability levels

7. Reevaluate the test, after it has been administered and graded, for guidance in reteaching, redesigning, and reconstructing, and for possible reuse.

Teachers who have followed such a basic outline are enthusiastic about its many long-range benefits. Students also have responded well

EVALUATING STUDENT PROGRESS

to test situations in which choices were offered, variety in types of question has been included, and questions have been arranged with simplest ones appearing at the beginning where they foster a feeling of confidence from the very outset. A good test is not easy to prepare, but if initially well-done and kept updated, it will give you several years of service as a valid evaluation instrument.

MARKING AND GRADING TEST PAPERS

The time and energy required for marking and grading test papers is inversely proportional to the time required for preparing the test and, generally, we must compensate for the time-consuming preparation of a comprehensive objective-type test by devising ways of scoring it with a minimum expenditure of time and energy.

Designing and Preparing Answer Sheets

It makes good sense to reduce all answers on an objective-type test to the form that can be recorded on an answer sheet (Figure 9-2) by marking designated slots corresponding to numbers or letters of student answers. A group of teachers at the Boonton High School has found that a further saving in time can be derived from the preparation of an accompanying scoring mask: they make one copy of a mimeographed answer sheet on heavier-weight paper and record the correct answers on that copy; then, using an IBM long-reach paper punch, they punch holes through those answer spaces to make a scoring mask. By placing the mask over student papers and lining up items properly, the correct answers can be viewed clearly through the punched holes, while incorrect responses, appearing as blanks, can be indicated with colored pen or pencil marks in the empty slots where correct responses should have appeared. The student's raw score on the test is calculated by subtracting the number of colored marks from the total possible number of points on the test.

Another enterprising teacher has devised a system by which she can "machine score" her objective tests. Using purple ditto masters, she prepares answer sheets in the usual way, and marks one copy with the correct answers. She then fastens the answer copy to a red carbon ditto master and draws circles around each correct answer so that only circles appear on the marking master thus prepared. When this master is used while running completed student answer sheets through a standard duplicating machine, red marks encircle student correct responses, and

BIOLOGY EXAMINATION ANSWER SHEET

MISS ALLEN _____ NAME
            _____ COURSE #
            _____ DATE

	1	2	3	4		1	2	3	4		1	2	3	4		1	2	3	4
1.	0	0	0	0	26.	0	0	0	0	51.	0	0	0	0	76.	0	0	0	0
2.	0	0	0	0	27.	0	0	0	0	52.	0	0	0	0	77.	0	0	0	0
3.	0	0	0	0	28.	0	0	0	0	53.	0	0	0	0	78.	0	0	0	0
4.	0	0	0	0	29.	0	0	0	0	54.	0	0	0	0	79.	0	0	0	0
5.	0	0	0	0	30.	0	0	0	0	55.	0	0	0	0	80.	0	0	0	0
6.	0	0	0	0	31.	0	0	0	0	56.	0	0	0	0	81.	0	0	0	0
7.	0	0	0	0	32.	0	0	0	0	57.	0	0	0	0	82.	0	0	0	0
8.	0	0	0	0	33.	0	0	0	0	58.	0	0	0	0	83.	0	0	0	0
9.	0	0	0	0	34.	0	0	0	0	59.	0	0	0	0	84.	0	0	0	0
10.	0	0	0	0	35.	0	0	0	0	60.	0	0	0	0	85.	0	0	0	0
11.	0	0	0	0	36.	0	0	0	0	61.	0	0	0	0	86.	0	0	0	0
12.	0	0	0	0	37.	0	0	0	0	62.	0	0	0	0	87.	0	0	0	0
13.	0	0	0	0	38.	0	0	0	0	63.	0	0	0	0	88.	0	0	0	0
14.	0	0	0	0	39.	0	0	0	0	64.	0	0	0	0	89.	0	0	0	0
15.	0	0	0	0	40.	0	0	0	0	65.	0	0	0	0	90.	0	0	0	0
16.	0	0	0	0	41.	0	0	0	0	66.	0	0	0	0	91.	0	0	0	0
17.	0	0	0	0	42.	0	0	0	0	67.	0	0	0	0	92.	0	0	0	0
18.	0	0	0	0	43.	0	0	0	0	68.	0	0	0	0	93.	0	0	0	0
19.	0	0	0	0	44.	0	0	0	0	69.	0	0	0	0	94.	0	0	0	0
20.	0	0	0	0	45.	0	0	0	0	70.	0	0	0	0	95.	0	0	0	0
21.	0	0	0	0	46.	0	0	0	0	71.	0	0	0	0	96.	0	0	0	0
22.	0	0	0	0	47.	0	0	0	0	72.	0	0	0	0	97.	0	0	0	0
23.	0	0	0	0	48.	0	0	0	0	73.	0	0	0	0	98.	0	0	0	0
24.	0	0	0	0	49.	0	0	0	0	74.	0	0	0	0	99.	0	0	0	0
25.	0	0	0	0	50.	0	0	0	0	75.	0	0	0	0	100.	0	0	0	0

Fig. 9-2 Answer sheet designed for quick and easy scoring of teacher-made, objective-type tests

empty circles readily identify those which are incorrect or missing. To avoid difficulties that might result from improperly lined-up answer sheets with the scoring master, she trial-runs a few unused answer sheets through the marking cycle to ascertain the correct positioning of circles and answer response slots. She then machine scores over 100 answer sheets in a matter of minutes.

Both scoring systems help to identify questions most frequently missed by students and to alert teachers to problem test items that should be examined more carefully to determine if remediation may be indicated or if the test items should be reworded, revised, or retired from use.

A Point Scoring System for Essay Answers

Grading essay answers is more demanding. Many teachers have reported satisfaction with the method of listing all factors to be included in a complete answer, and assigning differential point values to main factors, supporting statements, and examples cited. As they read each student paper, they record the point scores for each category, including points for additional contributing factors and for acceptable substitutes. The grade for the essay, assuming all other considerations to be acceptable, is then the total of the points scored, translated into a percentage value or a letter grade as specified by your grading system.

The values accruing from the use of this grading technique are practical and noteworthy: (1) it has been well-documented that when students understand the criteria upon which their answers will be evaluated, they develop their topics more fully and their work shows evidence of more careful thought and clear expression; and (2) teachers have experienced that the removal of essay answers from the realm of purely subjective considerations has made their position in grading more tenable—students seem more likely to concur with grades assigned to their work when specific factors are involved.

Student Participation in Grading

Generally, students perform better on graded work when they are familiar with the rating scale to be employed and when they have had an opportunity to participate in its determination. When given the responsibility of assigning point values to test questions, they appear to become very astute evaluators; they consider the degree of difficulty involved in answering questions requiring selection from a list of choices, interpreting information or data from an experiment, developing a biological concept, and deciding if a statement is true or false; and for each they are capable of making a supportable decision of point values to be earned, using a scale of 1 to 10. Having recognized the different levels of learning and accomplishment represented by these and other testing forms, they become more discerning when preparing answers. The result is inevitable; they learn more and at higher learning levels in order to earn higher scores.

There is also evidence to indicate that the quality and value of student projects and research papers improve as a result of student input into the determination of criteria upon which the work will be rated. In one case, a panel of judges was impressed by the unusual quality of individual student work they were asked to rate according to the students' own prepared list (Figure 9-3) of criteria upon which they requested that their work be judged.

RATING CARD FOR PROJECT NUMBER _____

Outstanding	= 10	Fair	= 4
Excellent	= 8	Poor	= 2
Good	= 6	Missing	= 0

OBJECTIVES.................................. _____
SCOPE OF STUDY........................... _____
METHOD OF APPROACH .... _____
PRESENTATION OF DATA .... _____
ANALYSIS OF DATA......................... _____
DISCUSSION OF RESULTS .... _____
CONCLUSIONS DRAWN..................... _____
SCIENTIFIC MERIT .... _____
ORGANIZATION & CLARITY .... _____
UNIQUENESS & CREATIVITY .... _____

               TOTAL .... _____

Fig. 9-3 Rating card designed by students lists criteria they request judges to consider when evaluating their projects

ENCOURAGING STUDENT SELF-EVALUATION

One of the best ways to help students become independent learners is to encourage them to approach a major study topic as a series of short-term goals which they themselves can evaluate. The success they experience at each level acts as a strong motivator for elevating their level of ambition and for the resetting of goals that involve higher levels of learning.

There are many techniques that can be employed for student self-

EVALUATING STUDENT PROGRESS

evaluation—and each is designed to serve a specific aspect of the learning program. However, the effectiveness of each can be enhanced by proper guidance and extended practice in use and by your innovativeness in developing adaptations and approaches that more personally suit your students or your teaching situation:

- Self-testing, using text guides and work sheets with questions arranged in sequence, assists students in self-evaluation and in identification of short term retention and reinforcement of learning of specific points.
- Programmed learning evaluation frames allow students to self-check a segment of work before attempting a more advanced level. Weaknesses are thus pinpointed for correction before continuing so that completion of any segment of work may be interpreted as mastery of the skills or successful completion of the stated objectives.
- Practice tests and review lessons in which students frame some of the questions help them to evaluate their mastery of a segment of work. Their questions may be used to quiz one another or may be submitted for possible inclusion on a test.

There is evidence which indicates that testing themselves and others is an effective way for students to reinforce their learning. In a comparison study involving two groups of evenly matched students, those in the group employing self-evaluation techniques scored over 20% higher on recall-type testing than did those in the group in which this technique was not particularly encouraged.

Not all situations call for precise answers. Often a question requires students to supply information and expression in the form of individually written essay answers. After writing a preliminary answer individually, students have been observed to profit by the experience of meeting in small groups for sharing ideas and deciding upon what elements comprise a complete answer. In the "give-and-take" atmosphere, they can sift out valuable and relevant points to be made and, on the basis of the interaction, each student can rewrite his essay. The experience is particularly revealing to individual students: each has an opportunity to assess the degree to which he contributed to the group effort and how well he was able to express and explain his ideas to others. In sharing, all learn, and more importantly, each learns to recognize his own strengths and limitations and to gain an insight into his own level of achievement and understanding.

Personal Assessment Inventories

Personal assessment inventories can be used periodically, generally before a testing period, to enable each student to check himself in

order to find out what more he needs to know and to plan a method by which the inadequacies can be remediated.
Example:

- Do you feel confident that you can analyze a food web illustrated by a chart of a biomass pyramid?
- Can you relate the nitrogen cycle to a generalized food web?
- Can you identify an herbivore, a carnivore, an omnivore, a parasite, and a saprophyte as they interact in the same food web?

Longer range assessments can also be planned:

- Occasionally ask students to list what they have achieved during a topic study. They are sometimes pleasantly surprised at how much they have learned.
- Occasionally interview students to assess and to help them assess progress being made in a given study under way. They have good insights into areas in need of improvement and often can be helped to overcome the weakness with some encouragement and guidance.
- At the end of a topic, ask students to assess how much they have learned by indicating individual achievements in a *before and after* study. One design that works well asks each student to indicate on a scale of 1 to 10 an assessment of his knowledge (B) before the topic was studied and (A) after its completion. Calculating the difference between A and B indicates his personal assessment of improvement for each item included, the total of which is his self-evaluation of learning and achievement, or of his accomplishment during the topic study.

A few examples taken from a test of 25 items on an inventory with reference to Study of Microscopic Forms of Life include:

Indicate how well you have the ability to:

1. use a microscope for study of microscopic organisms
   0 1 2 3 4 5 6 7 8 9 10   Difference = (5)
             (B)   (A)
2. distinguish between an amoeba and a paramecium
   0 1 2 3 4 5 6 7 8 9 10   Difference = (4)
                (B)   (A)

3. measure the length of a paramecium viewed at 100X
   0 1 2 3 4 5 6 7 8 9 10   Difference = (7)
   (B)   (A)
4. . . .
                                        TOTAL = __

DETERMINING STUDENT GRADES

While there are many methods by which student accomplishment on a particular segment of study can be evaluated, grades for an overall grading period must eventually be assigned. It is generally helpful to devise a list of criteria for defining and identifying the superior, the good, the average, the below average, and the failing student. With variations, as needed to suit a particular situation, the following list can be used to identify students in each category:

The student who qualifies for a grade of A:
- consistently engages in study and activity beyond that which is required
- demonstrates command and proper use of an appropriate biological vocabulary
- contributes significantly to almost every class and/or group discussion
- often initiates a discussion by introducing a well-phrased, appropriate, and relevant question or comment
- shows evidence of independent study in at least one major in-depth project study of biological significance
- demonstrates an unusually high degree of interest and involvement in all course-related activities
- demonstrates an ability to make associations and to rethink problems studied, with adaptation to new situations
- makes meaningful application of ideas and biological knowledge
- is enthusiastic and personally involved in all class projects via self-initiated as well as directed activities
- sets priorities and assumes full responsibility for the completion of all course work in a well-organized, neat, prompt, and thorough manner
- is alert and resourceful in the selection and use of proper references, resource materials, and laboratory equipment

- masters and practices proper use of appropriate laboratory skills
- has a critical and questioning attitude and often offers a valid alternative or challenge to an opinion or interpretation expressed by others
- demonstrates excellent insight into the long-range effects and significance as well as the social implications of applicable biological topics studied
- shows evidence of exceptionally-consistent high-quality performance on all tests, assignments, and laboratory reports

The student who qualifies for a grade of B:
- frequently engages in study and activity beyond that which is required
- is knowledgeable and expresses himself well with appropriate employment of biological terminology
- makes worthwhile contributions to at least 75% of class and/or group discussions
- has completed a major project or has engaged in an independent study project
- demonstrates active involvement and genuine interest in and/or concern for classroom activities
- consistently produces better-than-average work with an evenness in quality that does not vary significantly by more than one letter grade
- is prompt, neat, thorough, and usually accurate (by at least 85%) in all work
- is conscientious and gives careful attention to all assignments
- usually accepts full responsibility for all phases of course work
- is receptive to constructive criticism of work and/or approach and takes positive action to improve
- applies general biological principles studied to new situations
- demonstrates good understanding of biological topics studied and manipulates biological concepts with ease
- exhibits an ability to work effectively in the laboratory and to employ appropriate skills with dexterity

The student who qualifies for a grade of C:
- completes all required work in a satisfactory manner and acceptable fashion—according to directions and stated standards and specifications

- expresses himself moderately well with biological terms
- actively participates in at least 50% of all class and/or group discussions
- has successfully completed a required research paper or an assigned project
- is attentive to all class activities and exhibits mild interest in the proceedings
- prepares all assignments carefully, neatly, and in accordance with the specified time schedule
- responds favorably to corrections and suggestions for improvement and seeks assistance or otherwise takes steps to upgrade the quality of work produced
- produces work that is acceptable in quality, with some unevenness in performance which may vary by one or two letter grades
- sometimes probes for reasoning behind statements or solutions offered and occasionally suggests valid opinions as alternates
- exhibits an acceptable level of understanding of biological concepts but experiences some difficulty with their application
- partially accepts responsibility for his own work and achievement
- is adequate in the use of appropriate laboratory skills

The student who qualifies for a grade of D:
- usually makes some attempt to complete work which is required
- occasionally participates in class discussions with worthwhile contributions
- exhibits evidence of some research or project work
- has a languid interest in most class activities and allows his attention to wander on occasion
- sometimes appears to be passive and uninterested in the course
- rarely questions or displays a critical or questioning attitude
- exhibits a poor understanding of the biological topics studied beyond an attempt to master some of the factual material
- displays considerable inaccuracy in mastery of the factual material
- frequently "misunderstands" assignments or is in other ways unprepared for class
- fails to plan, prepare, and complete assigned work carefully, neatly, and in an orderly fashion

- appears to be willing but is slower than average in complying with directions, instructions, and remedial procedures
- fails to assume a personal responsibility for his own success and for seeking assistance when the need for such is indicated
- produces work that shows an occasional unevenness in quality but, on balance, is below average
- is deficient in his ability to employ proper manipulatory skills in the laboratory

The student who qualifies for a grade of F:

- fails to complete the required quality and/or quantity of course work
- is listless and inattentive to class proceedings
- seldom participates in class discussions
- shows no evidence of engaging in research and/or project work
- is easily distracted and often becomes involved in matters unrelated to the classroom activity or course
- rarely asks questions or probes into the meaning or application of topics investigated
- refuses to accept responsibility for his own accomplishment
- is usually unprepared or late with assignments
- is careless in work habits and disregards directions given
- has a poor understanding of biological topics studied and retains only fragments of general principles of the course
- is unable to function satisfactorily in laboratory studies and investigations

It is important to keep abreast of the progress being made by individual students and to discuss the matter with them well in advance of the end of a grading period. Students do not always have a realistic view of the over-all quality of their work—often they count on one good quiz grade to overshadow many areas of weakness or neglect. It should be helpful to discuss with them the descriptions of A, B, C, D, and F graded students to allow each one to see where he fits and if, indeed, that is where he wants to be at the end of the grading period. Hopefully, together you can work out a plan for up-grading the work of those who are not satisfied with their recognized standing.

In the case of unsatisfactory performance, additional action should be taken. Some teachers at the Boonton High School have found that advising the guidance counselors and notifying the parents via a telephone call or a written report, as shown in Figure 9-4, helps to draw

BOONTON HIGH SCHOOL
335-9700
INTERIM REPORT OF UNSATISFACTORY WORK

RE: Student _____ Teacher _____

Course _____ Grade _____

Homeroom _____ Counselor _____

Dear Parent:

This report is being sent to inform you that your son/daughter is doing unsatisfactory work in biology. While it does not necessarily indicate a failure at this time, parent cooperation is being sought to encourage the student to make prompt and significant improvement in the specific areas that have been identified as trouble spots.

_____ Poor test performance   _____ Excessive absence from class

_____ Lack of participation in   _____ Lack of preparation for
      class activities   class/laboratory

_____ Failure to complete   _____ Careless study and work
      assignments   habits

_____ Failure to report for extra   _____ Failure to follow up on lab
      help   investigations

Additional comments: _____

ACTION REQUESTED

_____ Student see teacher for assistance during the helping class period from 2:15 - 2:45 p.m.

_____ Parent call teacher to discuss this matter

_____ Parent contact guidance counselor to schedule a student-parent-teacher-counselor conference

Additional recommendations: _____

_____

_____

This matter has been discussed with the student and his guidance counselor has been advised of this parent contact.

_____   _____   _____
Student signature   Date   Teacher signature

Parent's name: _____

Address: _____

Fig. 9-4 Format used for notifying parents of unsatisfactory student work and/or progress

attention to the problem and to urge that corrective action be taken before the potential failure becomes a reality.

EVALUATION OF THE COURSE

Teacher accountability and teacher professionalism dictate that teachers should constantly assess and evaluate their professional performance. Active involvement in continuing-education and in-service programs and in constant revisions of courses of study attest to their dedication to the attainment of this goal. In addition, an evaluation program has implications for evaluation of the teaching as well as the learning that takes place, and evaluation of the course offerings as viewed by students as well as by the teacher, who bears the major responsibility for its success. For this, student evaluations can be helpful and should be openly solicited from former students who can now view a course with perspective, and from students who have just completed the course for the more immediate reactions.

What students think about a course does matter. It will determine, to a degree, the enrollment in the course and the demand for advanced and/or independent study courses as a follow-up of a first-year biology course. It is particularly revealing if you can see a course as students perceive it to be.

Course evaluation sheets can be helpful for this purpose. Students may be asked questions relating to the most and the least liked topics, the use of teaching assists, individual vs. group work activities, and general attitude toward laboratory, library, and field work. Often, it is not the topic but the manner of presentation that is most beneficial (or devastating) to a student. The results of student evaluation sheets, coupled with performance records, may indicate that a change is needed.

Questions commonly used for course evaluation by students include:

- What was the *one* strongest aspect of the biology course?
- What was the *one* weakest aspect of the biology course?
- Was the experiment on *enzyme activity* of value in meeting the course objectives?
- What was your reaction to the filmstrip on *genetic engineering*?
- What did you like most about the course?
- What did you like least about the course?
- What *one* change would you like to see made in the course content?

- What one change would you like to see made in the presentation of the course?

An effective evaluation program is a constructive process; it carefully monitors the progress of individual student learning programs; it measures the quality and quantity of work produced by individual students in terms of what they have achieved; and it determines how well the course offerings and teaching methods facilitate the attainment of student goals and their individual performance at optimum levels. Through identification of areas of ineffectiveness, appropriate action can be taken to direct the evaluation program toward achieving its ultimate goal—the improvement of the performance level of students in a quality biology education.

10 Planning and Maintaining a Laboratory Program

The laboratory is the focal center of instruction in a high school biology course which emphasizes the inquiry approach to the study of life. It is here that students can be encouraged to become active participants in absorbing studies which promote the application of scientific processes to numerous problems associated with life and which provide opportunities for extensions of open-ended investigations and self-initiated activities. It is here that biology can be most effectively presented as a dynamic study of life.

The laboratory should be more than a physically well-equipped, well-organized workroom. It should also create an atmosphere that is conducive to activity and involvement of all students; it should feature opportunities for the study of life and the kinship of all of its various forms; it should accommodate some spontaneous as well as planned activities; it should enable students to become increasingly adept at performing technical skills and employing appropriate methods of approach to studies of biological problems; and its program and format of activities should become progressively less structured, allowing students greater opportunity and encouragement for self-directed study. Consequently, via their lab activities, students should experience a personal growth in resourcefulness and independence in learning biological knowledge while developing a concept of the methods used in biological science in the continuing search for understanding of the nature of life and of the problems associated with it.

PLANNING LABORATORY ACTIVITIES

Laboratory experiences can do much to dispel the feeling that biology lies primarily within the realm of encyclopedic learning, and that its study involves the memorization of a collection of facts. But lab activities, too, must be carefully scrutinized; to ensure that the lab sessions serve a useful purpose of discovery and do not represent mere "exercises" in which students engage, activities being considered for inclusion in the program should be screened for their suitability, in accordance with certain critical factors:

Checklist for a Laboratory Investigation

1. Does it involve a biological principle or concept?
2. Does it involve a problem or situation concerning life?
3. Does its study involve living specimens or a life situation?
4. Does it encompass the broad purposes and goals of the course?
5. Does it allow the student to engage in the investigation on his own?
6. Does it allow for every student to be an active participant?
7. Does it require a minimum of technical skill and material, in accordance with the student's resources and maturity level?
8. Does it allow students to employ scientific methods in the lab (careful observations, precise measurements, collection and analysis of data)?
9. Does it encourage students to form their own conclusions about the significance of the data and of the investigation?
10. Does it have student interest and appeal?
11. Does it have meaning and relevance to the student in terms of the topic under consideration?
12. Does it allow for the primary study to be extended beyond the planned scope?
13. Does it have integrity as an *investigation* in that it poses a problem for solution that is not described in the textbook?
14. Does it have simplicity of design that offers students a reasonable assurance that they will succeed at discovery?
15. Does it have order that proceeds according to a sequence that students can follow?

LIVING SPECIMENS IN THE LABORATORY

Investigative studies in biology clearly involve life and life situations for which it is important that living specimens be provided. Since there is usually some existing familiarity with small animals, investigations involving their growth rates, behavioral responses, and genetic patterns make interesting and satisfying studies for beginning students. Hence, practice in the basic elements of the scientific method can be afforded by activities which also contribute to an overall grasp of a problem under consideration and to the development of important values in the form of interest and respect for life and life forms. There are, of course, restrictions that must be placed on studies involving small animals in the laboratory, and students should be made aware that some of the experimental studies they read about cannot and must not be attempted by anyone other than a highly experienced and professional researcher engaged in an authorized project.

Proper maintenance and use of small animals can be stated in two simple but all-inclusive rules which summarize the guidelines offered by the Animal Welfare Institute:

1. Animals being observed by students must be maintained in optimum conditions of health, comfort, and well-being.
2. No vertebrate animal used in high school teaching may be subjected to any experimental procedure which interferes with its normal health or which causes it pain or distress of any kind.

Suitable animals can be obtained from reputable suppliers or may be donated for short-term study by students. However, only specimens in good health should be considered, and at no time should those which are potentially dangerous or to which there are known student allergies be accepted. Adequate and appropriate housing must be provided, and regimens for care and feeding must be set up; students must be trained for responsibilities of feeding and cleaning, and for "adoptive care" during weekend and school vacation periods.

While it is not advisable to allow all students to actually handle small animals, all should know and understand the proper procedures. One teacher posts rules in a prominent position above each small animal cage:

FOR THE SAFETY AND WELL-BEING
OF THIS ANIMAL AND YOURSELF

1. Refrain from giving the animal a feeling of falling
2. Refrain from making loud or unnecessary noises
3. Refrain from making sudden or fast movements
4. Refrain from using forcible restraint when holding the animal

She reports that students enthusiastically explain to visitors the chain reaction that might be set into motion if a rule is broken: if the animal appears to be falling (breaking rule #1), someone may scream (breaking rule #2 causes the animal to feel threatened), one or more individuals may suddenly reach out in an attempt to rescue it (breaking rule #3 causes the animal to associate fast movement with a possible attack), and the person who catches it may hold it in a tight grip (breaking rule #4 causes the animal to feel restrained and he will react). In the process someone may be scratched or bitten, and the animal may suffer a bad fright that may have a harmful and permanent effect on his behavior and disposition.

The technique has proved to be very successful; she reports that, since posting the rules and enlisting the cooperation of her students, there has not been a single mishap involving any animal or student.

Providing Suitable Specimens for the Teaching Situation

A wide assortment of plants and animals cultured and maintained in the laboratory heightens the interest level of students and makes possible some unplanned investigations. The availability of a fur-bearing animal, for example, makes possible an on-the-spot comparison of human vs. animal hair or claw vs. nail structure; aquarium specimens may provide for a spontaneous determination of the rate of operculum movements or an analysis of tail and fin activity in swimming; and an array of terrarium specimens may help to clarify the relationship that exists between sporophyte and gametophyte generations of a moss plant. In some cases, individual projects may be suggested by the activities of animals in the lab, and students may be motivated to pursue studies involving their behavioral, reproductive, and genetic patterns.

The primary purpose of the living organisms is, of course, for planned activities, and efforts should be made to obtain specimens which are the most suitable. Those which are hardy, dependable in their performance, and versatile in use are usually the most practical—after some experimentation you can determine the right specimen "mix" for your courses. The energy expended in this effort will pay handsome dividends in (1) savings in time, energy, and money, and (2) a more realistic student concept that all organisms engage in the full spectrum of life activities and are not limited to the one aspect of life for which they may have been singled out for a specific study.

In a teaching situation, the well-known species of protozoa, bacteria, fungi, algae, and selected "standard" invertebrates and members of higher plant and animal groupings are needed because much resource material has been amassed from extensive studies about them. But new materials are also needed to effect a change-of-pace and, in some cases, to obtain more suitable material. It is important to exercise good judgment when selecting living material for use in the laboratory.

Guidelines for Selecting Living Materials

1. Select only those specimens for which there is a need and a specific use. Plants and animals should not be maintained for display or decorative value alone.
2. Select only those specimens which are easily maintained and for which there are adequate provisions in the form of housing and proper care.
3. Select organisms which are versatile in use. For example, Drosophila can be used in genetic studies, chromatographic analyses, metamorphosis and life history studies, investigations of cell physiology, and investigations into the effects of diet on sperm motility.
4. Do not allow sick or infected plants or animals in the laboratory. This could cause infection of other organisms. Sometimes it becomes necessary to remind students that the laboratory is neither a plant nor an animal hospital.
5. Select fast-growing, easily-reproduced plants (Coleus, Tradescantia), those grown quickly from seed (bean, pea, squash, zinnia, marigold), useful water plants (duckweed, algae, Elodea), and animals that lend themselves to laboratory culture (protozoa, lower invertebrates, crickets, snails, small fishes, etc.).
6. Experiment with some alternates as possible replacements or substitutes for traditional organisms that may have been overused: Xenopus laevis, the African clawed frog, is a good substitute for Rana pipiens for embryological and physiological studies; blow fly (Lucilia seratica), flesh fly (Sarcophaga bullata), and black fly (Simulium vittatum) larvae are larger and easier to dissect for salivary gland chromosome study than are Drosophila species; and Physarum polycephalum, a slime mold, is easily activated from sclerotial to plasmodial form for determining the rate of its cytoplasmic streaming—and with the added advantage that it is easier to culture and is more dependable than are amoebae.

STUDENT INVOLVEMENT IN THE "LIVING LABORATORY"

Students should assist in the growing of cultures and in the maintenance of terraria, aquaria, growing chambers in the classroom/laboratory or greenhouse, and the animal room. A well-organized program can result in benefits to both teacher and student participants; students can learn a great deal about the life forms they are caring for and the teacher is afforded an opportunity to set priorities for activities in the laboratory. By allowing your students to assist with the routine maintenance chores, you can more easily concentrate your time and energy in areas of planning and providing lab experiences which foster student inquiry, creativity, and imagination.

Training Student Laboratory Assistants

Advanced students who are highly motivated and exhibit an aptitude for laboratory work should be trained as assistants. While programs vary widely, those with which teachers express the greatest satisfaction have certain elements in common: they set high standards of eligibility for the position; they place great importance on the assignment; they offer the student assistant a real opportunity to learn and grow, while assuming limited responsibility in the lab; and the appointments are looked upon by all students as highly-prized and coveted opportunities. By not permitting the assignment to be diminished by low standards, nor the areas of responsibility limited to clean-up tasks, prestige and status are accorded the position of "student lab assistant," to which serious-minded first-year students can aspire.

The relationship between teacher and lab assistant is something akin to that mentor/protégé. They may work together on some preparations, or the teacher may program some training materials, in order that the assistant may learn and employ new techniques and methods. The teacher, however, must also be supportive of the assistant's aspirations and constantly encourage him to pursue investigations on his own, willingly make suggestions, offer guidance, and delegate responsibility to the assistant as he exhibits increasing competence and readiness.

The assistant/first-year student relationship is also profitable; students seem to be able to communicate more openly with others of closer age and experience, and the accessibility of an additional resource person in the lab is both a boon to the beginning student and an opportunity for the assistant to reinforce his own learning and understanding of basic knowledge and its application.

A student lab assistant, selected by careful screening, is bound to be a capable person who, properly trained, will perform satisfactorily. If his duties are planned to be diversified and charged with satisfying responsibilities, his presence in the laboratory will be a real help to the busy teacher.

Let's look at some suitable activities for a laboratory assistant:

1. Maintain bacteria culture collection according to the protocol of your established method.
2. Cultivate protozoa, yeast, and various algae and invertebrates to be used in laboratory investigations. There are many culture methods which he can check out and experiment with in search of the most satisfactory method for the conditions and requirements for your lab.
3. Assist first-year students in setting up experiments and in assembling materials and equipment needed for investigations.
4. Assist students with the mechanics involved in laboratory procedures: focusing a microscope, making a wet mount for microscope viewing, constructing a manometer for use in respiratory rate studies, using chemical indicators, etc.
5. Prepare staining solutions and chemical indicators and reagents to be used in the laboratory. (Examples: Wright's stain, Lugol's iodine solution, methylene blue stain, etc.)
6. Prepare and stain specimens for laboratory viewing.

    Example: PROCEDURE FOR VITAL STAINING OF PROTOZOA: To a specified volume of protozoan culture in a centrifuge tube add a weak solution of the desired vital stain and spin down in a centrifuge until organisms have gravitated to the bottom of the tube; pour off supernatant and resuspend stained organisms in fresh media for viewing.

7. Prepare solutions of desired concentrations, for laboratory use.

    Examples: MOLAR SOLUTION: 1 gram molecular weight of a substance dissolved in 1 liter of distilled water.

    $$0.25M\ (NH_2)_2CS = N - (2 \times 14) = 28$$
    $$(Thiourea)\quad H - (4 \times 1) = 4$$
    $$C - (1 \times 12) = 12$$
    $$S - (1 \times 32) = \underline{32}$$
    $$76$$
    $$76 \times .25 = 19\ g.$$

    19 g. $(NH_2)_2CS$ dissolved in 1 liter of H_2O produces the 0.25M Thiourea solution.

    NORMAL SOLUTION: 1 gram equivalent weight of dissolved substance in 1 liter of distilled water (1 gram equivalent weight is the molecular weight of the substance divided by the number of positive or negative ionic charges).

$$0.1N\ NaOH = H - (1 \times 1) = 1$$
$$Na - (1 \times 23) = 23$$
$$O - (1 \times 16) = \underline{16}$$
$$40 \times 0.1 = 4\ g.$$

4 g. NaOH dissolved in 1 liter of H_2O produces the 0.1N NaOH solution. Had the substance been H_2SO_4 with 2 positive ionic charges, a further division by 2 would have been necessary following the determination of 0.1 times its molecular weight.

PPM (parts per million): 1 ppm is 1 gram of a substance dissolved in 1,000,000 ml. (or 1,000 liters) of water.

RINGER'S SOLUTIONS:

	Mammalian	Amphibian	Drosophila
NaCl	9.00 g.	6.50 g.	7.50 g.
KCl	0.42 g.	0.14 g.	0.35 g.
$CaCl_2$	0.24 g.	0.12 g.	0.21 g.
$NaHCO_3$	0.20 g.	0.20 g.	—
H_2O	1 liter	1 liter	1 liter

0.85% PHYSIOLOGICAL SALINE: Dissolve 8.5 g. NaCl in 1 liter of distilled water.

8. Set up apparatus for demonstrations to be performed. (Examples: osmosis, diffusion, root pressure, transpiration, chromatography.)

PROVIDING NEW AND INNOVATIVE MATERIALS AND ACTIVITIES

With or without a laboratory assistant, new methods and materials need to be explored. Efforts must be made to keep the laboratory up-to-date, efficient, well-equipped, well-organized, and operating at a minimum cost in terms of time, energy, and money expended, while still supplying fully the needs of a quality laboratory program.

Investigating New Procedures and Methods

The effectiveness and suitability of new and different procedures for use in the laboratory and for preparation of materials to be used in the teaching-learning situation should be checked. Some teachers for whom established routine methods were something less than ideal have found satisfaction with new methods tried and adopted:

- Plastic petri dishes can be sterilized for reuse by treating with solutions of sodium hypochlorite.
- Anhydrous silica gel provides for long-term maintenance and storage of fungal and bacterial cultures. Not only does this method extend the storage time to a year or more, but with fewer transfers necessary there is a lessening of the danger of their contamination and they may remain useful for several years.
- Silicone culture gum prolongs the viewing time of wet-mount microscope slides. The silicone allows for diffusion of O_2 and CO_2 and retards evaporation so that the same material can be viewed at intervals over a period of several days, or made available to students who missed a laboratory observation due to absence.
- Nigrosin stain (10%), normally used as a relief stain for definition of ciliary and outline structures, is also useful for facilitating the viewing of contractile vacuole formation and ciliary action in paramecium feeding. Dramatic effects have been achieved when specimens have simultaneously been stained with Nigrosin and fed congo red-stained yeast—as digestion occurs within the vacuoles, the biochemical reactions can be detected via a red-to-blue color change.
- Calcozine Magenta XX is less expensive, easier to use, and more stable with age than the aceto-carmine dye usually used for onion root cell chromosome studies.
- Giemsa stain is an excellent stain for root tip cell chromosome studies using a variety of plant specimens. Excellent results for squash root cells as well as for onion have been reported.
- Incense is a good producer of smoke particles for Brownian Motion demonstrations and for use in comparative studies of molecular motion vs. motility of microorganisms.

- Finquel, a poikilothermic anesthetic, is desirable for preparing organisms for physiological studies—specimens not used can be quickly and easily revived.

Investigating Versatile and Excellent-Performing Living Specimens

Sometimes we are apt to overlook the excellence of commonly known materials for use in the laboratory. However, if we examine the many uses to which some easily obtained materials can be put, we see some excellent teaching tools: the hen's egg, for example, can be used to demonstrate osmosis and plasmolysis, in addition to providing material for embryological studies; the succulent house plant *Kalanchoe daigremontiana* (Bryophyllum) demonstrates vegetative propagation, tropisms, photosynthesis, effects of hormones on photoperiodism, and the allelopathic influence of the parent plant on plantlets which fall from the leaf margin to the soil close to the parent; and the common bread mold may be used for studies of reproductive patterns and in studies of both the enhancing and the inhibitory influences of environmental factors on growth rates and patterns. Probiotic as well as antibiotic concepts can be developed in these latter studies.

Some not-so-well-known specimens also provide impressive material for study:

- *Strongylocentrolus purpuratus*, a sea urchin, is an excellent specimen for observation of the gamete release-fertilization-cleavage sequence and the effects of temperature and pH on the time interval between the stages.
- *Pilobolus crystallinus*, the so-called "shot-gun fungus," demonstrates the phototropic response in a very dramatic fashion.

Setting up Work Centers in the Laboratory

Supervision of laboratory work requires more teacher time and preparation than does a regular recitation-type lesson. Productive laboratory sessions do not just happen; rather, they are more often the result of foresight and planning on the part of the teacher.

Organizing work centers for specific laboratory procedures can enhance the efficiency with which your students perform and can facilitate the effectiveness with which you supervise their activity. For example, one workbench, equipped with a sink area, wall clock with a second sweep hand, pegboard for hanging staining racks, shelves for holding staining materials, and closets for storing staining supplies can become a Microscope Slide Staining Center. A sign frame mounted on the wall above the work counter and a portfolio of placards on which are printed the protocol for staining procedures make excellent additions to this work area; the student has but to select the desired placard from the portfolio and insert it into the frame to effect a handy guide and reference for his work. Each staining procedure should be outlined simply, in a step-by-step fashion as shown below:

PROTOCOL FOR THE GRAM STAIN

1. Prepare and fix bacterial smear for staining.
2. Cover smear with Hucker's ammonium oxalate crystal violet stain and allow to remain for 1 to 2 minutes.
3. Rinse with water.
4. Flood film with Gram's iodine solution and allow to act for 1 minute.
5. Rinse with water.
6. Decolorize with 95% alcohol (30 seconds to 1 minute, depending on the thickness of the smear) until alcohol draining from the slide is no longer tinged with color.
7. Rinse with water.
8. Counterstain with safranin for about 10 to 30 seconds.
9. Rinse with water and blot dry between pages of a bibulous paper book.
10. Examine stained smear using an oil immersion lens and determine the gram morphology of the organism.

The convenience of the technique is matched by its usefulness in keeping confusion and mishaps to a minimum; employment of the proper staining solutions for the proper time periods and in the proper sequence is usually achieved without the need to introduce additional references to the staining center.

Other areas can also be designated for other specialized procedures where the use of appropriate "lab-aids" can be introduced. Using colored ink, labels on reagent bottles can be marked to indicate the purpose for which each is useful, and labels on indicator bottles can be marked similarly with the effective pH range of the indicators. It has been my personal experience that techniques such as these contribute significantly to the success with which diverse activities can be engaged in simultaneously in the laboratory, and to the effectiveness with which individual student work can be supervised.

Time and Labor-Saving Procedures in the Laboratory

There are many routines that can be set up for efficiency in the laboratory. Some may involve your students, and you should enlist their cooperation and input into maintenance procedures.

When microscope slides are used in series—as with embryological stages of development in the chick—slides should be color coded, with a different color marking on the label of each stage in the series. This helps students to locate slides and return them to their proper storage places, saves sorting time otherwise demanded of you or your lab assistant, and helps to keep an accurate accounting of materials.

Keeping your supply company catalogs readily accessible in a central location makes for easier reference and greater convenience when preparing requisitions and orders.

Maintaining a WANT LIST on clipboard, tablet, or order sheet posted on the supply room door will enable you to keep a record of supplies as they are running low or as the last unit is put into use. This will ensure that you never run out of a needed material.

Keeping a desk blotter-size calendar, with large enough day blocks for entering designated arrival dates for living materials from supply companies, will enable you to make appropriate plans and have everything in readiness for their arrival, to plan precisely to the day when and how they will be used, and to synchronize all other plans for materials needed to accompany the specimens in the investigation or activity.

Setting up a schedule for microscope cleaning and servicing, refrigerator cleaning and defrosting, culture transfers, and other routine lab-maintenance chores will eliminate the danger that all will require attention at the same time or that some will be neglected or overlooked.

Placing laminated copies of care and operating instructions for each piece of equipment (incubator, spectrophotometer, autoclave, centrifuge, etc.) on the back or side of the instrument will make them readily available for use. (NOTE: It is recommended that you keep the original or a copy of each on file, for insurance against loss.)

Keeping a file of instructions, warranties, serial numbers of pieces of equipment, and record of location (storage, current usage, or interdepartmental loan) will save time and embarrassment if there is a need for documents, an inventory, or an accounting.

Keeping a file of bulletins and leaflets distributed by leading biological supply houses is an invaluable aid for reference about uses of stains, care of plants, culture methods for protozoa and invertebrates, marine and fresh-water aquaria maintenance, etc. This file serves doubly well if cross-referenced.

Being selective in the ordering of equipment and supplies has long-range as well as immediate benefits: the best for the purpose and those which are multi-purpose in use are usually the most economical. For example, coplin staining jars are versatile in use, serving as developing chambers for thin layer chromatography performed on microscope slides as well as for their intended use as staining jars. Their double-duty use helps to alleviate some storage and maintenance problems.

Keeping a card file of names, addresses, and telephone numbers of repair and service agencies, suppliers of special materials such as fertile hens' eggs, and representatives who will demonstrate new models and equipment saves much time and energy that would otherwise be required when the information is needed.

Keeping a card file for culture methods, stain preparations, formulas for culture media, plant and animal care, etc., makes information easily available for use. The cards, placed in glassine envelopes, are not damaged when used, and if they are color-coded as well, they can be easily located by subject area and returned to proper positions in the file box.

SAFETY IN THE LABORATORY

There is nothing that is more important in the laboratory than the safety of your students. Bright light, heat, chemicals, glass, fumes, instruments, electricity, living organisms, an open flame, even air and water—all necessary elements in a biology laboratory—can become potentially hazardous if improperly handled. Students must be instructed in safety measures to be taken and constantly reminded of their responsibilities for knowing and practicing procedures that contribute to safety in the laboratory setting.

Eye-catching cartoons and slogans on safety posters placed in strategic positions in the laboratory do much to raise the safety consciousness of students. Ultimately, however, the responsibility for operating a safe laboratory lies with the teacher. Here we can make no compromise nor allow any concession which is at variance with well-thought-out rules designed for the safety of students; we must govern this aspect of the laboratory program with a strong hand.

Safety Rules for Students (Figure 10-1) for general laboratory procedures should be included in the student biology handbook, and a full lab session should be devoted to the lab, its safety features, and the reasons for the rules set forth. Later, for specialized work, such as microbiological studies or investigations involving radioactive materials, additional supplements can be provided. Helping students to develop a safety-consciousness that is always in evidence requires constant vigilance. Students must be alerted at the beginning and throughout lab sessions in which there may be a specific hazard, and reminders must be given repeatedly for safety measures to be practiced until they become habit. Then, as new and more specialized techniques and materials are introduced, the accompanying safety measures must be stressed until they, too, become incorporated into the students' repertoire.

Perhaps the single most effective influence on the development and practice of safe laboratory techniques by students is our own personal example via employment of good—and safe—techniques. Wearing a laboratory coat or jacket and employing other protective devices when applicable, putting into practice safety-oriented techniques in all demonstrations, and providing and using facilities with built-in safety features all contribute to a good image and a model for students to emulate.

Safety in the laboratory can be further enhanced by many specific precautions and practices:

- Be observant of how students use equipment and assist them in skillful and safe handling of all materials.
- Provide automatic pipettors, rubber bulbs for use on pipettes, or cotton-plugged mouthpieces on pipettes.
- If mirrors are to be used for reflected light use in microscope viewing, warn students against the use of direct sunlight.
- Provide individual, disposable, sterile equipment for personal use (such as lancets for blood-letting, toothpicks for scraping epithelial cheek cells, mouthpieces for spirometer, etc.) and guard against their reuse.
- Guard against the removal of lancets, hypodermic needles, and syringes from the laboratory. Count and account for the return of every item distributed or made available for use.
- Provide for safe disposal of waste chemicals, radioactive materials and broken glassware, and provide for autoclaving of microbiological cultures and used media.
- Provide a zephirin or other disinfecting solution for soaking

SAFETY RULES FOR BIOLOGY STUDENTS

Never handle materials or equipment in the laboratory without specific directions from the teacher.

When carrying laboratory equipment or materials, give this activity your undivided attention. Do not try to do something else at the same time.

Be very careful when working with a flame, with glassware, with chemicals, with electricity, with instruments, and with organisms.

Handle all specimens carefully and according to instructions.

Report any accident, spill, or personal injury immediately to the teacher.

Report any allergies you may have so that special adjustments to laboratory studies can be made for you.

Use safety goggles for experiments requiring bright or UV light, acids, or any other hazardous situation.

Report any unusual occurrence — broken glass, peculiar odor, water leak, etc. . . . — immediately to the teacher.

Always read labels on containers and always label substances that you are working with; never use an unlabeled substance.

Never bring food or gum into the laboratory.

Keep long hair out of danger by making good use of an elastic band when working in the laboratory.

Do not lean back on chairs or stools. Protect yourself from the possibility of falling by keeping all chair legs flat on the floor.

Always be courteous and considerate of others and of their work.

Always use materials and equipment with care and according to instructions.

Always read, listen to, and follow directions.

Refer to appropriate directions for proper handling and techniques before starting work with any specialized materials.

Know the location of fire extinguisher, eye wash station, spill control center, and fire blanket for use in an emergency.

Always use good judgment and practice rules for safety in the lab.

REFER TO THIS PAGE FREQUENTLY FOR INFORMATION ABOUT THE LABORATORY.

Fig. 10-1 SAFETY RULES FOR BIOLOGY STUDENTS is a handy reference for general laboratory procedures when included in Guidelines for Biology Students

used pipettes and contaminated glassware and utensils prior to their washing.
- Practice careful supervision in physiological studies such as bradycardia or cardiobrachial responses involving student subjects.
- Avoid the use of toxic materials whenever possible; provide vacuum-packaged specimens treated with odorless, non-toxic preservatives and immobilize small living specimens by lowering the temperature or by using club soda instead of chemical anesthetics.
- Make certain that all safety devices (fire extinguisher, fume hood, fire blanket, eye wash station, and spill control center) are operational.
- Have a direct line of communication with the school nurse for speedy response to emergency calls.
- Never culture pathogens in the laboratory, but insist that your students treat all cultures as if they were pathogenic.
- Provide and display RADIOACTIVE warning labels, tags, and tape for appropriate use.
- Provide protective rubber gloves, asbestos mitts, protective goggles, rubber bands, and lab coats or aprons (for staining sessions an old shirt is a good cover-up) for use when needed.
- Provide proper storage facilities for volatile, radioactive, and other potentially hazardous or controlled substances and enforce rules governing their proper use and storage.

LABORATORY WORK AND COMMUNITY RELATIONS

Communications with the public are important, particularly in relation to laboratory work where some unfavorable publicity—most often incorrect—is of concern to parents and other interested parties. Parents are genuinely interested in what their children are doing in school, and their activities within the biology laboratory setting provide excellent opportunities for keeping parents informed in an interesting way. For example, during an open-house session, students can run a spirometer testing service for parents and visitors—demonstrating the techniques for the determination and interpretation of individual results, with complete explanation of the significance of any existing discrepancies between expected (calculated) and actual performance.

Or, a muscle fatigue determination, using an Ergograph attachment with a Kymograph apparatus, can be demonstrated, with students interpreting individual visitors' graph tracings resulting from their personal involvement in the demonstration.

Parents should also be advised of any unusual activities involving students. The typing of blood, used in genetic studies and other investigations involving blood chemistry, should be announced to parents (see Figure 10-2) with assurances of safety precautions to be taken against infection.

Boonton High School
306 Lathrop Avenue
Boonton, New Jersey 07005
(201)-335-9700

Principal **Vice Principal**
Robert Kane **Michael Renaldo**

Dear Parent:

We are planning to conduct student blood typing tests in Biology class on _____ . Although there is no obligation, it might prove to be helpful and informative if all biology students participate in the testing program. Please be assured that all precautions to ensure aseptic conditions will be practiced throughout the procedure. If you would like to give your permission for your son/daughter to take part in the testing program, please sign in the space provided below and return this permission slip.

Sincerely,

D. Allen

Dorothea Allen
Department of Biology

_____   _____
       (Student Name)   (Parent Signature)

Fig. 10-2 Communicating with parents by letter keeps them informed about unusual laboratory activities involving students

Students enjoy and profit from laboratory experiences which are designed to allow them to make discoveries in a hands-on situation. When they can share with their parents and friends, they reinforce their learning with new enthusiasm, and when a new area for study opens up new problems, there is encouragement for further investigation. As the laboratory approach diminishes the teacher-dominated atmosphere, the students learn to be resourceful and to discover, while at the same time they discover to learn. It is a highly-desirable outcome and the ultimate to be achieved in high school biology courses—it is a reaffirmation of the laboratory as the central mechanism for learning.

Recommended Reading and Reference

Ann Arbor Science for the People Editorial Collective (eds.), *Biology as a Social Weapon*. Minneapolis, Minn.: Burgess Publishing Company, 1977.

Agel, Jerome, *Is Today Tomorrow? A Synergistic Collage of Alternative Futures*. New York: Ballantine Books, 1972.

Altman, Philip L. and Dorothy S. Dittmer (eds.), *Biology Data Book*, 2nd ed., Vol. I. Washington, D.C.: Federation of American Societies for Experimental Biology, 1972.

———, *Biology Data Book*, 2nd ed., Vol. II. Washington, D.C.: Federation of American Societies for Experimental Biology, 1973.

———, *Biology Data Book*, 2nd ed., Vol. III. Washington, D.C.: Federation of American Societies for Experimental Biology, 1974.

Anderson, Hans O. and Paul G. Koutnik, *Toward More Effective Science Instruction in Secondary Education*. New York: The Macmillan Company, 1972.

Barber, Bernard, "The Ethics of Experimentation with Human Subjects," *Scientific American* 234:2, 1976.

Bedwell, Lance E., "Developing Environmental Education Games," *The American Biology Teacher* 39:3, 1977.

Bingman, Richard M., *Learning Through Inquiry—A Social Design*. Kansas City, Missouri: Mid-Continental Regional Educational Laboratory, 1970.

Brophy, Jere E. and M. Everton, *Learning From Teaching: A Developmental Perspective*. Rockleigh, N.J.: Allyn and Bacon, 1976.

Brown, D. Frank, *Education By Appointment*. West Nyack, N.Y.: Parker Publishing Co., Inc., 1978.

Causton, David R., *A Biologist's Mathematics*. Baltimore, Maryland: University Park Press, 1977.

Deaton, John G., *Below the Belt*. Palisade, N.J.: Franklin Publishing Company, Inc., 1978.

———, *New Parts for Old*. Palisade, N.J.: Franklin Publishing Company, Inc., 1974.

Eckholm, Eric P., *Losing Ground; Environmental Stress and World Food Prospects*. New York: W.W. Norton Company, 1976.

Ehrlich, P. R. and R. L. Harriman, *How to Be a Survivor; A Plan to Save Space Ship Earth*. New York: Ballantine Books, 1971.

Epstein, Sam, *Scientific Instruments*. Palisade, N.J.: Franklin Publishing Company, Inc., 1970.

Franke and Franke, *Man and the Changing Environment*. New York: Holt, Rinehart, and Winston, 1975.

Fuller, W. (ed.), *The Biological Revolution: Social Good or Social Evil?* Garden City, Long Island: Doubleday & Company, 1971.

Galley, F. B. (ed.), K. Petrusewicz, and L. Ryskowski, *Small Mammals: Their Productivity and Population*. New York: Cambridge University Press, 1975.

Gans, Carl, *Biomechanics*. Philadelphia, Pa.: J. B. Lippincott Company, 1974.

Givner and Graubard, *A Handbook of Behavior Modification for the Classroom*. New York: Holt, Rinehart, and Winston, 1974.

Goldstein, Kenneth, *The World of Tomorrow*. New York: McGraw Hill Book Co., 1969.

Gordon, Thomas, *TET, Teacher Effectiveness Training*. New York: Peter H. Wyden, 1976.

Gray, C. H. (ed.-in-chief), *Laboratory Handbook of Toxic Agents*. Palisade, N.J.: Franklin Publishing Company, Inc., 1968.

Grobman, Arnold B. (ed.), *Social Implications of Biological Education*. Reston, Va.: NABT Publications, 1976.

Hardin, Garrett, "Living on a Lifeboat," *BioScience* 10:561, October 1974.

Harding, Volker, and Fagle, *Creative Biology Teaching*. Ames, Iowa: University of Iowa Press, 1969.

Hechlinger, Adelaide, *Biochemistry Units for the High School Biology Teacher*. West Nyack, N.Y.: Parker Publishing Co., Inc., 1973.

———, *Modern Science Dictionary*. Palisade, N.J.: Franklin Publishing Company, Inc., 1975.

Hellman, Hal, *Population*. Philadelphia, Pa.: J. B. Lippincott Company, 1972.

Howard, Ted and Jeremy Rifkin, *Who Should Play God?* New York: Dell Publishing Co., Inc., 1978.

Kibler, Robert J. et al, *Objectives for Instruction and Evaluation*. Rockleigh, N.J.: Allyn and Bacon, 1974.

Klinckman, Evelyn (BSCS Supervisor), *Biology Teacher's Handbook*, 2nd ed. New York: John Wiley & Sons, 1970.

Knobloch, Irving Williams, *Readings in Biological Science*, 3rd ed. New York: Irvington Publications, Appleton-Century-Crofts, 1973.

Krebs, Charles J., *Biology: The Experimental Analysis of Distribution and Abundance*. New York: Harper and Row, 1978.

Kurtz, P., "The Uses and Abuses of Science," *The Humanist* 23:9, 1970.

La Palme, Norman, *Metric Workshop*. Palisade, N.J.: Franklin Publishing Company, Inc., 1975.

Lee, Addison E. and H. Edwin Steiner, Jr., "The Research Potential of Inquiry Objectives," *The American Biology Teacher* 32:9, 1970.

Lewis, James, *Administering the Individualized Instruction Program*. Englewood Cliffs, N.J.: Prentice-Hall Inc., 1971.

Maddox, John, *The Doomsday Syndrome: An Attack on Pessimism*. New York: McGraw Hill Book Co., 1972.

Marshall, J. and L. Hales, *Classroom Test Construction*. Reading, Mass.: Addison-Wesley Publishing Company, 1971.

Mehrens and Lehmann, *Measurement and Evaluation in Education and Psychology*, 2nd ed. New York: Holt, Rinehart, and Winston, 1978.

Morholt, Evelyn, Paul Brandwein and Alexander Joseph, *A Sourcebook for the Biological Sciences*. New York: Harcourt, Brace, and World, 1966.

Moment, Gairdner B. and Helen M. Habermann, *Biology: A Full Spectrum*. Baltimore, Md.: Williams and Watkins, 1973.

———, *Mainstreams of Biology*. New York: Oxford University Press, 1977.

Moyer, Wayne A., *Biophysical Science Activities for the High School*. West Nyack, N.Y.: Parker Publishing Co., Inc., 1972.

NABT, *Compendium of Reprints on the Theory of Evolution and of the Evolution-Creationism Equal Time Controversy*. Reston, Va.: NABT, 1977.

Newman, Thelma R., *Changeover . . . Breakthrough to Individualism*. Wayne, N.J.: Wayne Township Public Schools, 1976.

Newton, David E., *Biological Science Casebook*. Portland, Maine: J. Weston Walch, Publisher.

Norman, *Dimensions of the Future: Alternatives for Tomorrow*. New York: Holt, Rinehart, and Winston, 1974.

Odum, *Ecology: The Link Between the Natural and the Social Sciences*. New York: Holt, Rinehart, and Winston, 1975.

Orlans, F. Barbara, *Animal Care from Protozoa to Small Mammals*. Reading, Mass.: Addison-Wesley Publishing Company, 1977.

Orr, Howard and John C. Marshall, *Introduction to Computer Programming for Biological Scientists*. Boston, Mass.: Allyn and Bacon, 1974.

Perlman, Philip, *General Laboratory Techniques*. Palisade, N.J.: Franklin Publishing Company, Inc., 1964.

Phillips, Butt, and Metzger, *Communication in Education: A Rhetoric of Schooling and Learning*. New York: Holt, Rinehart, and Winston, 1974.

Porzio, Ralph, *The Transplant Age*. New York: Vantage Press, Inc., 1969.

Quagliarello, E., F. Palmieri and Thomas Singer (eds.), *Horizons in Biochemistry and Biophysics*, Vol. 4. Reading, Mass.: Addison-Wesley Publishing Company, 1977.

Ray, James D. Jr. and Gideon E. Nelson, *What a Piece of Work is Man*. Boston, Mass.: Little, Brown, and Co., 1971.

Roberts, Arthur D., *Educational Innovation: Alternatives in Curriculum and Instruction*. Boston, Mass.: Allyn and Bacon, 1975.

Rosengren, John H., *Biology Teacher's Guide*. West Nyack, N.Y.: Parker Publishing Co., Inc., 1968.

Rosenzweig, Michael L., *And Replenish the Earth: The Evolution, Consequences, and Prevention of Overpopulation*. New York: Harper and Row, 1974.

Schatz, Vivian and Albert Schatz, "One Puff of Death," *The American Biology Teacher* 34:4, 1972.

Shebesta, Donald F., "Substituted Sammy: An Exercise in Defining Life," *The American Biology Teacher* 34:5, 1972.

Snowbarger, Barbara C., "New Perspectives for Viewing Inquiry," *The American Biology Teacher* 32:5, 1970.

Stronck, David R., "The Scientific Revolution in Science Teaching," *The American Biology Teacher* 33:6, 1971.

Swanson, Carl P., *Natural History of Man*. Englewood Cliffs, N.J.: Prentice-Hall, Inc., 1973.

Truitt, Willis H. and T. W. Graham Solomons, *Science, Technology, and Freedom*. Boston, Mass.: Houghton-Mifflin Company, 1974.

Unruh and Alexander, *Innovations in Secondary Education*. New York: Holt, Rinehart, and Winston, 1974.

Van Vleck, David B., *The Crucial Generation: Your Challenges and Your Choices*. Charlotte, Vermont: Optimum Population, Inc., 1973.

Ward's *Stain Preparation and General Usage*, Rochester, N.Y.: Ward's Natural Science Establishment.

Williams, Byram L. and Keith Wilson (eds.), *A Biologist's Guide to Principles and Techniques of Practical Biochemistry*. Baltimore, Md.: University Park Press.

Wilson, E. O., *Sociobiology: The New Synethesis*. Cambridge, Mass.: Harvard University Press, 1975.

APPENDICES

A

The Metric System

GENERAL SYSTEM OF MULTIPLES

Multiple	Prefix	Symbol
10^{12}	tera	T
10^{9}	giga	G
10^{6}	mega	M
10^{3}	kilo	k
10^{2}	hecto	h
10	deca	da
10^{-1}	deci	d
10^{-2}	centi	c
10^{-3}	milli	m
10^{-6}	micro	μ
10^{-9}	nano	n
10^{-12}	pico	p
10^{-15}	femto	f
10^{-18}	atto	a

LINEAR MEASURE

10 millimeters = 1 centimeter
10 centimeters = 1 decimeter
10 decimeters = 1 meter
10 meters = 1 decameter
10 decameters = 1 hectometer
10 hectometers = 1 kilometer

LIQUID MEASURE

10 milliliters = 1 centiliter
10 centiliters = 1 deciliter
10 deciliters = 1 liter
10 liters = 1 decaliter
10 decaliters = 1 hectoliter
10 hectoliters = 1 kiloliter
1000 milliliters = 1 liter

SQUARE MEASURE

100 sq. millimeters = 1 sq. centimeter
100 sq. centimeters = 1 sq. decimeter
100 sq. decimeters = 1 sq. meter
100 sq. meters = 1 sq. decameter
100 sq. decameters = 1 sq. hectometer
100 sq. hectometers = 1 sq. kilometer

WEIGHTS

10 milligrams = 1 centigram
10 centigrams = 1 decigram
10 decigrams = 1 gram
10 grams = 1 decagram
10 decagrams = 1 hectogram
10 hectograms = 1 kilogram

B

Acid-Base Indicators Used in Biology

Name of Indicator	Concentration recommended (% dye in 50% ethyl alcohol)	pH Range	Color Change
Thymol blue (acid range)	0.04	1.2 – 2.8	Red to yellow
Brom phenol blue	0.04	3.0 – 4.6	Yellow to blue
Congo red	0.10	3.0 – 5.0	Blue to red
Methyl orange	0.05	3.0 – 4.4	Orange to yellow
Brom cresol green	0.04	3.8 – 5.4	Yellow to blue
Methyl red	0.02	4.4 – 6.4	Red to yellow
Chlor phenol red	0.04	4.8 – 6.4	Yellow to red
Brom cresol purple	0.04	5.2 – 6.8	Yellow to purple
Brom thymol blue	0.04	6.0 – 7.6	Yellow to blue
Phenol red	0.02	6.8 – 8.4	Yellow to red
Cresol red	0.02	7.2 – 8.8	Yellow to red
Thymol blue (alkaline range)	0.04	8.0 – 9.6	Yellow to blue
Phenolphthalein	0.05	8.3 – 10.0	Colorless to red
Alizarin yellow	0.10	10.0 – 12.0	Colorless to yellow

C

Biological Stains and Their Use

NAME OF STAIN	USE
Acetocarmine	Chromosome stain in squash preparations
Aniline Blue	Cytoplasmic stain; counterstain for nuclear stain
Carbol Fuchsin	Bacterial spore stain; fungi stain
Crystal Violet	Bacterial stain; primary Gram stain
Congo Red	Vital stain
Eosin B; Eosin Y	Counterstain for blue nuclear stain
Fast Green B	Counterstain for botanical tissue
Fuchsin	Nuclear stain
Gentian Violet	Bacterial stain
Hematoxylin	General tissue stain
Indigo Carmine	Bone and cartilage stain
Janus Green	Vital stain; mitochondria stain
Malachite Green	Botanical stain; bacterial stain
May-Greenwald Stain	Blood stain
Methylene Blue	Vital stain; bacterial stain; blood stain
Neutral Red	Vital stain; cytoplasmic granule stain
Nigrosin	Bacterial spore stain; cell content stain
Orcein	Nuclear stain; mitotic figures stain
Phloxine B	Cytoplasmic stain
Safranin	General botanical stain; Gram counterstain
Sudan Black	General counterstain
Sudan III; Sudan IV	Fat stain
Toluidine Blue	Vital stain; cartilage stain in whole mounts
Trypan Blue	Vital stain; connective tissue stain
Wright's Stain	Blood stain

For further reference:

Ward's *Stain Preparation and General Usage*, from which the above data were abstracted, with the permission of Ward's Natural Science Establishment, Rochester, N.Y. 14903.

D

Temperature Characteristics of Homoiothermic Animals

ANIMAL	NORMAL RECTAL TEMPERATURE (in °C)	TEMPERATURE REGULATING MECHANISMS*		
		Sweating	Shivering	Panting
Man	37	+	+	−
Camel	37–38	+	+	−
Cat	39	−	+	+
Cattle (dairy)	38–39	+	...	+
Chicken	41–42	−	+	+
Dog	38–39	−	+	+
Donkey	36–38	+	+	...
Guinea pig	39	−	+	+
Horse	38	+	...	...
Monkey	37–38	+	+	−
Mouse (white)	37	−	...	−
Ostrich	38	...	...	+
Pigeon	43	−	+	+
Rabbit	39	−	+	+
Rat (white)	37–38	−	+	−
Seal (fur)	38	+	+	+
Sheep	39–40	+	+	+
Swine	37–38	−	+	+

* + = present; − = absent

For further information:

Philip L. Altman and Dorothy S. Dittmer (eds.), *Biology Data Book*, 2nd ed., Vol. II. Bethesda, Maryland: Federation of American Societies for Experimental Biology, 1973. Table 95, p. 863, from which the above data were abstracted, with the permission of the editors.

E

Animal Reproduction

SPECIES	AGE* AT SEXUAL MATURITY	GESTATION PERIOD* (IN DAYS)	YOUNG NUMBER* PER LITTER
Cat (Felis catus)	(6–15 mo)	63	4
Chipmunk (Tamias striatus)	(2.5–3 mo)	31	(3–6)
Cow (Bos taurus)	(6–10 mo)	284	1
Dog (Canis familiaris)	(6–8 mo)	151	(1–22)
Donkey (Equus asinus)	1 yr	365	1
Elephant (Elephas maximus)	(8–16 yr)	624	1
Goat (Capra hircus)	8 mo	151	(1–5)
Gorilla (Gorilla gorilla)	♀ (5–8 yr)	270	1
Guinea pig (Cavia porcellus)	(55–70 da)	68	(1–8)
Hog (Sus scrofa)	(5–8 mo)	(101–103)	(6–15)
Horse (Equus caballus)	1 yr	336	1
Man (Homo sapiens)	♀ (11–16 yr)	(253–303)	1
Mink (Mustela vison)	1 yr	(39–76)	(4–10)
Mouse (Mus musculus)	35 da	(19–31)	(1–12)
Muskrat (Ondatra zibethicus)	1 yr	(19–42)	(1–11)
Opossum (Didelphis marsupialis)	♀ 6 mo	(12–13)	(5–13)

*Values in parentheses are ranges.

SPECIES	AGE* AT SEXUAL MATURITY	GESTATION PERIOD* (IN DAYS)	YOUNG NUMBER* PER LITTER
Rat (Rattus norvegicus)	(40–60 da)	21	(6–9)
Sheep (Ovis aries)	(7–8 mo)	(144–152)	(1–4)
Squirrel (Sciurus carolinensis)	(1–2 yr)	44	(1–6)
Whale (Balaenoptera physalus)	3 yr	360	1

For further information:

Philip L. Altman and Dorothy S. Dittmer, (eds.), *Biology Data Book*, 2nd ed., Vol. I. Bethesda, Maryland: Federation of American Societies for Experimental Biology, 1972. Table 13, parts 1 and 2, pp. 137–138, from which the above data were abstracted, with permission of the editors.

F

Chromosome Number in Animal Species

SPECIES	DIPLOID NUMBER OF CHROMOSOMES/SOMATIC (BODY) CELL
Asterias forbesi (starfish)	36
Cambarus clarkii (crayfish)	200
Canis familiaris (dog)	78
Cavia porcellus (guinea pig)	64
Daphnia magma (daphnia)	20
Drosophila melanogaster (fruitfly)	8
Felis catus (cat)	38
Homo sapiens (man)	46
Hydra vulgaris (hydra)	32
Lumbricus terrestris (earthworm)	36
Macaca mulatto (Rhesus monkey)	42
Mus musculus (mouse)	40
Musca domestica (housefly)	12
Pan troglodytes (chimpanzee)	48
Periplaneta americana (cockroach)	34
Planaria torva (planarian)	16
Rana pipiens (frog)	26
Rattus norvegicus (rat)	42
Scypha ciliata (sponge)	26
Xenopus laevis (African clawed frog)	36

For further information:

Philip L. Altman and Dorothy S. Dittmer (eds.), *Biology Data Book*, 2nd ed., Vol. I. Bethesda, Maryland: Federation of American Societies for Experimental Biology, 1972. Table 1, Parts 1 and 2, pp. 1–7, from which the above data were abstracted, with the permission of the editors.

G

Gene Expression in Humans (reliably reported)

TRAIT	DOMINANT GENE EXPRESSION (DD or D–)	RECESSIVE GENE EXPRESSION (rr)
Digits on fingers and toes	extra digits	normal number
Dimpled cheeks	present	absent
Earlobes	free	attached
Eyelashes	long	short
Finger length	short, stubby	normal
Hair color	dark	light
Hairline	widow's peak	straight
Hair form	curly	straight
Hair whorl	clockwise	counterclockwise
Long palmar muscle	absent	present
Mental ability	normal	feebleminded
Mid-digital hairs	present	absent
Palate	cleft with harelip	normal
Skin pigment	normal pigmentation	albino
Taste ability	PTC taster	PTC non-taster
Toe length	2nd toe longest	big toe longest
Toe arrangement	webbed	separated
Tongue musculature control	can fold and/or roll	cannot fold and/or roll
Tooth enamel	defective, dark, or absent	normal, present
Upper lateral incisors	absent	normal, present

H

Gene Distribution in World Populations

DISTRIBUTION OF BLOOD GROUP ALLELES IN WORLD POPULATIONS*

POPULATION	ABO BLOOD GROUPS			
	O	A	B	AB
USA, White population	46.0%	40.0%	10.0%	4.0%
USA, Black population	49.3%	26.0%	21.0%	3.7%
Swedish	37.9%	46.7%	10.3%	5.1%
Japanese	31.2%	38.4%	21.8%	8.6%
Chinese	30.0%	25.0%	35.0%	10.0%
Hawaiian	36.5%	60.8%	2.2%	0.5%
Australian aborigine	53.1%	44.8%	2.1%	0.0%
North American Indian	91.3%	7.7%	1.0%	0.0%

*reliably reported

FREQUENCY OF PRESENCE OF LONG PALMAR MUSCLE IN ETHNIC GROUPS AND OTHER PRIMATES*

Ethnic Group	Percentage of individuals possessing recessive genotype (rr) for presence of Palmaris Longus muscle
Chinese	98
Japanese	97
USA, Black population	95
Russians	87
USA, White population	86
Polish	81
French	75
Other Primates	
Gibbons	100
Chimpanzees	95
Gorillas	15

*reliably reported

I

Dilution Table for Liquids

Percentage strength of original liquid

		100	96	95	90	85	80	75	70	60	50	40	30	20	15	10	8	5	4	3
	95	5	1																	
	90	10	6	5																
	85	15	11	10	5															
	80	20	16	15	10	5														
	75	25	21	20	15	10	5													
Percentage strength of liquid required *and* Volumes of original liquid to be used	70	30	26	25	20	15	10	5												
	60	40	36	35	30	25	20	15	10											
	50	50	46	45	40	35	30	25	20	10										
	40	60	56	55	50	45	40	35	30	20	10									
	30	70	66	65	60	55	50	45	40	30	20	10								
	20	80	76	75	70	65	60	55	50	40	30	20	10							
	15	85	81	80	75	70	65	60	55	45	35	25	15	5						
	10	90	86	85	80	75	70	65	60	50	40	30	20	10	5					
	8	92	88	87	82	77	72	67	62	52	42	31	22	20	7	2				
	5	95	91	90	85	80	75	70	65	55	45	35	25	15	10	5	3			
	4	96	92	91	86	81	76	71	66	56	46	36	26	16	11	6	4	1		
	3	97	93	92	87	82	77	72	67	57	47	37	27	17	12	7	5	2	1	
	1	99	95	94	89	84	79	74	69	59	49	39	29	19	14	9	7	4	3	2

Volumes of diluent to be added

Reprinted from *Ward's Ready Reference Lab Guide*, a Curriculum Aid, courtesy of Ward's Natural Science Establishment, Rochester, N.Y. and with their permission.

INDEX

A

Ability levels:
 advanced and gifted, 98
 career opportunities, 98–99
 college-bound, 97–98
 slow learner, 95–97
Abiogenesis, 142
Aceto-carmine dye, 211
Achievement, 82–83
Activity type questions, 185
Advanced level courses:
 biology, 75–77
 guidelines for developing, 74–75
 placement program, 78–79
Advanced students, 98
After-image, 117
Agenda for day, 54
Alternate forms of tests, 180
Amino acids, DNA coding, 123
Amoeba, mercury, 113–114, 146–147
Anecdote, dramatic, 52–53
Anhydrous silica gel, 211
Animals, 205–206
Answer sheets, 189–191
Antibiotics, sensitivity of bacterial species, 112–113
Articulation of chicken foot, 117
Assessment inventories, 193–195
Assignments:
 extra, 44

Assignments (cont.)
 guide sheets, 54, 56
 long-range, 46
 options, 93–95
 personalizing, 89–90
 special, 45
Assistants, student laboratory, 208–210
Atherosclerosis, 167
Atmosphere, 18–22
Attitudes:
 interpreting data, 143–144
 predictions, 144–146
 research project, 100
Autotutorial program:
 avoid pitfalls, 69–70
 how to write, 68–70
 individualized *rates* of learning, 85
 mastery of subject matter content, 85
 potential, 66
 self-pacing, 85

B

Bacteria, sensitivity to U.V. light, 111
Bacterial species, sensitivity to antibiotics, 112–113
Bactericidal action, lysozyme, 111
Balance, 48–50
Beating heart, 147
Bedwell, Lance E., 40–42
Bingo-type games, 38

Biochemical and biophysical patterns:
  effect of scale on organisms, 134–135
  mode of action of penicillin, 133–134
  physiology of blood clot, 132–133
Biochemical-biophysical topics, 135–136
Bioelectric wave patterns, 176
Bioethical consciousness, 158–160
Biogenesis vs. Abiogenesis, 142
Biology:
  advanced course, 75–77
  as a process, 51
  career opportunities, 98–99
  general education, 140
  mathematics, 121–131 (see also Mathematics)
  synchronizing studies with other classes, 135–136
Biology Baseball, 37
Biology Bingo, 38
Biology Book of Records, 90–92
Biology Football, 37
Biology student's handbook, 43–44
Biomonitor, 166–167
Blind spot, 116
Blood clot, 132–133
Blood counts, 125–126
Blood pressure, 167
Botanical Treasure Hunt, 25
Branching form, 68
Bromelin, protein digesting action, 113
Brownian Motion, 211
BSCS pamphlets, 47, 60
Bud development, 116
Bulletins, 215

C

Caffeine, 172
Calcozine Magenta XX, 211
Calendar, desk blotter-size, 214
Carbon, pathway in photosynthesis, 175
Card files, 215
Cardiac muscle activity, 167
Career opportunities, 98–99
Cartridged 8mm film, 71
Catalogs, supply, 214
Cause of malaria, 142
Cell division, 122–123
Cellular respiration, 42

Change of pace, 51–52
Charts:
  clipped, laminated, 71
  data, 56
Chi Square test, 129–131
Cholesterol build-up, 167
Chromatographic analysis of leaf pigments, 176
Chromatographic separations of biological substances, 119
Ciliary streaming, 114
Circulation, fish tail, 114
Classifications, 53
Classroom reference center, 59–61 (see also Reference center)
Classroom resources, 45
Cloning, 157
Closed circuit T.V., 108–109
Coleus blumei leaves, 175
Coliform bacteria, 174
Collections, 71
College-bound student, 97–98
Color coding, slides, 214
Communication:
  individualized study options, 96
  verbal, 44
Community relations, 218–220
Conclusions, 204
Conduction, plant tissues, 115
Contraction of skeletal muscle fibers, 115
Contracts:
  options, 73
  using, 102–104
Contractual fulfillments, 180
CO_2 gas produced by fermentation, 117, 118
Course:
  description and topic outline, 43
  evaluation, 200–201
Crayfish, internal functions, 114–115
Creation, 157
Credit for outside work, 100–101
Crossword puzzles, 38
Curiosity, 22–26
Cytoplasmic streaming, 117

D

Data:
  organizing, 56

Data (cont.)
  skills for interpreting, 143–144
Data-chart association, 56
Death, 147–149 (see also Social/moral issues)
Debates, 150–151
Decision making, 29–30
Demonstrations:
  broad preparation, 106
  closed circuit T.V., 108–109
  display, 115–116
    conduction in plant tissues, 115
    stages in bud development, 116
    stages in frog embryology, 116
  dramatic focal point, 109
  follow-up quiz, 107
  handout sheets, 107
  introduction, 107
  lean explanations, 109
  long-range, 120
  motivate interest, 107
  on-going, 120
  practice, 106
  professionalism, 106–108
  projecting or taping, 113–115
    ciliary streaming, 114
    circulation in fish tail, 114
    contraction of skeletal muscle fibers, 115
    feeding response in hydra, 114
    internal functions of crayfish, 114–115
    mercury amoeba, 113–114
    osmosis, 115
  researching topic, 106
  responsibility for preparations, 107
  short, simple design, 107, 109
  student input, 107
  student involvement, 116–119
    absorption of CO_2 gas produced by fermentation, 117–118
    after-image, 117
    all students, 116–118
    articulation of chicken foot, 117
    CO_2 production during fermentation, 118
    cytoplasmic streaming, 117
    epidemic, 118–119
    eye dominance, 117
    "hole-in-the-hand," 117
    location of blind spot, 116
    preparation, 119

Demonstrations (cont.)
  student involvement (cont.)
    volunteers, 118–119
  that work, 109–111
    bactericidal action of lysozyme, 111
    O_2 production during photosynthesis, 110
    oxidative enzyme in fresh produce, 110–111
    source of oxygen breathed by fish, 109
    transpiration by plant leaves, 109–110
  time, 109
  time-lapse, 111–113
    phototropic response in euglena, 112
    protein digesting action of bromelin, 113
    sensitivity of bacterial species to antibiotics, 112–113
    sensitivity of bacteria to U.V. light, 111–112
  topic selection, 106
  varied techniques, 109
  vehicle, 107–108
Description, course, 43
Dialogues, 181
Diastolic Pressure, 167
Diffusion through membrane, 119
Discussions:
  controversial issues, 151
  small groups, 181
  student-centered, 56
Display:
  demonstrations for, 115–116
  motivation, 26
Dissolved oxygen in water, 169
DNA, extraction, 119
DNA coding, amino acids, 123
DNA research, 141
Doubling of cells in cell division, 122–123
Drosophila crosses, 128–129, 130, 177
Drugs, 172

E

Education, general, 140
Electrophoretic analysis of leaf pigments, 176
Electrophoretic separations of biological substances, 119
Embryology, frog, 116

Emerson enhancement effect in
  photosynthesis, 119
EMI Programmed Learning Booklets, 60
Enrichment experiences, 152
Epidemic, 118–119
Ergograph, 168
Escherichia coli, 165
Essay questions, 183–184
Essay-type answer, 182, 191
Ethics, 158–160
Ethics boards, 158
Euglena, phototropic response, 112
Evaluation:
  alternate forms of tests, 180
  constructing examinations and unit
    tests, 187–189
  contractual fulfillments, 180
  course, 200–201
  daily quizzes, 180
  designing program, 179–181
  dialogues, 181
  discussions with small groups, 181
  games used for, 39–40
  grades, determining, 195–200
    A, 195–196
    B, 196
    C, 196–197
    D, 197–198
    F, 198
  guide sheets, 54
  individual lab reports, 180
  individual reports, 180
  innovative strategies, 183–187
    activity type questions, 185
    essay questions, 183–184
    false statements, 184
    games, 183
    interpretations of biological topics,
      184–185
    presenting problem for students to
      solve, 187
    questions based on graph analysis of
      data, 185–187
    35mm slides, 187
  marking and grading tests, 189–192
    designing and preparing answer
      sheets, 189–191
    essay answers, 191
    point scoring system, 191
    student participation, 191–192
  one-to-one encounters, 181

Evaluation (*cont.*)
  open book and/or notebook tests, 181
  pretests, 180–181
  report of unsatisfactory work, 199–200
  self, 192–195
    personal assessment inventories,
      193–195
    practice tests and review lessons, 193
    programmed learning frames, 193
    text guides and work sheets, 193
  small group interactions, 180
  standardized tests, 181
  student involvement, 30–31
  teacher-made tests and quizzes, 181
  values of opposing forms, 181–183
    factual vs. interpretative, 181–182
    oral vs. written, 182–183
    short answer vs. essay-type answer,
      182
Evolution, 157
Examinations (*see* Evaluation)
Experiment kits, 71
Eye dominance, 117

F

Factual evaluation, 181–182
False statements, 184
Feedback, immediate, 66
Feeding response, hydra, 114
Fermentation, 117, 118
Field study, 24
Files, 214, 215
Film:
  cartridged 8mm, 71
  guide sheets, 56
Find the Fallacy, 37
Finquel, 212
Fish, source of oxygen, 109
Fish tail, circulation, 114
Follow-up, 96–97
Format:
  biology paper, 62–63
  written reports, 43
Frog embryology, 116

G

Games:
  Bingo-type, 38–39

INDEX 245

Games (cont.)
  Biology Baseball, 37
  Biology Bingo, 38
  Biology Football, 37
  board, 40–42
  Cellular Respiration, 42
  crossword puzzles, 38
  evaluation, 39–40, 183
  extend learning, 38–39
  Find the Fallacy, 37
  Hangman, 37
  HERBO, 38
  hidden word puzzles, 38
  Identify the Imposter, 40
  MANNO, 38
  Natural Selection, 37
  Photosynthesis, 42
  Plant Identification, 37
  Population Genetics, 42
  review, 38
  "scrambled words," 38, 39
  Spell-Down, 37
  stimulate learning, 37–38
  teacher-prepared, 37
  word, 38
  word hunt, 38
  ZOOLO, 38
Geiger-Mueller tube, 175, 176
Genetic engineering, 157
Genetics Bingo, 39
Genetics studies, probability in, 126–129
Giemsa stain, 211
Gifted students, 98
Grading, 43, 103, 189–192, 195–200
Gram stain, 213
Graph analysis, questions based on, 185
Graphs, 71
Great Experiments in Biology, 79
Group interaction:
  asking questions, 57–58
  evaluating comments, 58
  expressions of opinion, 56
  guidelines for leading, 58–59
  respect, 56
  responsibility for statements, 57
  self-expression, 56
  sharing information, 58
  size of group, 56
  student-centered discussion, 56
Guidelines for Biology Students, 44–46
Guide sheets, 54, 56

H

Handbook, biology student's, 43–44
Handout sheets, 71, 107
Hangman, 37
Hardin, Garrett, 158
Heart, beating, 147
Heart attacks, 167
Hematocrit, 176
HERBO, 38
Hidden word puzzles, 38
Hippocratic Oath, 149
"Hole-in-the-hand," 117
Human reproduction, 156–157
Hydra, feeding response, 114
Hypertension, 167

I

Identify the Imposter, 40
Illustrative materials, 70–72
IMPScope, 176
Incense, 211
Independent study, 44, 66, 100
Individual enrichment experiences, 152
Individualized learning, 47
Individualized study options:
  ability levels, 95–99
    advanced and gifted students, 98
    career opportunities, 98–99
    college-bound student, 97–98
    slow learner, 95–97
  attitude, 100
  autotutorial methods, 85
  communication, 96
  contracts, 102–104
  credit for outside work, 100–101
  development of model, 86
  differences among students, 85
  evaluation, 86
  follow-up, 96–97
  independent study and research, 100
  interest and support, 100
  investigation, 96
  learning activities, 86
  mastery of content, 85
  materials and resources, 100
  motivation, 86, 96, 100
  open classroom, 101–102

Individualized study options (cont.)
  personalized learning program, 86–92
    compiling *Biology Book of Records*, 90–92
    developing a scenario, 87–89
    personal interest, 88
    personalizing assignments, 89–90
  phases of learning process, 86
  planning, 86
  providing, 92–95
    assignments, 93–95
    setting priorities, 92–93
  rates of learning, 85
  readiness, 100
  reinforcement of learning, 86, 96
  self-pacing, 85
  student-centered program, 85
  subject matter content, 86
  time for research, 100
  variations provided, 86
Individual lab reports, 180
Individual reports, 180
Information, sharing, 58
Innovation:
  advanced level courses, 73–79
    biology, 75–77
    developing, 74–75
    placement program, 78–79
  cartridged 8mm film, 71
  collections, 71
  evaluation, 183–187
  experiment kits, 71
  handout sheets, 71
  illustrative materials, 70–72
  involvement and achievement, 82–83
  laboratory, 210–213
  living specimens, 71
  microscope slides, 71
  minicourses, 79–82
    benefits, 81–82
    full semester course, 81–82
    microbiology, 80–81
    starting, 80–81
  pictures, charts, graphs, 71
  plastic embedded specimens, 71
  programmed learning, 66–70
    autotutorial, 66, 68–70
    avoid pitfalls, 69–70
    branching form, 68
    linear form, 67–68

Innovation (cont.)
  programmed learning (cont.)
    predictive form, 68
    writing unit, 69
  reading materials, 71
  responsibility for learning, 72–73
    active involvement, 72
    conferring with students individually, 73
    goals and objectives, 72
    learning on contractual basis, 73
    open classroom/laboratory atmosphere, 73
    open-ended studies, 72
    options or free-choices, 72
    self-pacing, 72
    small group discussions, 73
    value sessions, 73
Instructions, care and operating, 214
Instruments and techniques:
  biochemical techniques, 168–171
    dissolved oxygen in water, 169–170
    Thunberg, 170–171
  bioinstrumentation, 162–168
  Biomonitor, 167
  Ergograph, 168
  Kymograph, 167–168
  lung capacity and spirometer, 165–166
  microscope, 162–164
  multi-technological approach, 171–177
    metabolism, 171–174
    pathway of carbon in photosynthesis, 175–176
    water pollution, 174–175
  Pneumograph, 168
  spectrophotometer, 164–165
  sphygmomanometer, 166–167
Integrity of investigation, 204
Interaction, group, 56–59
Interest:
  increasing, 17–42 (*see also* Motivation)
  maintaining high level, 50–51
Interpretations of biological topics, 184–185
Interpretative evaluations, 181–182
Interpreting data, skills, 143–144
Interviews, 150
Inventories, personal assessment, 193–195
Investigation, 96
Involvement, student, 28–31, 82–83

INDEX

Issues, 52, 141–160 (see also Social/moral issues)

K

Kits, experiment, 71
Koch's postulates, 119
Kymograph, 167–168

L

Lab-Aids, 71
Laboratory:
  living, 23–24
  safety, 43
Laboratory assistants, student, 208–210
Laboratory program:
  active participation of student, 204
  community relations, 218–220
  integrity of investigation, 204
  interest and appeal, 204
  living specimens, 204–207, 212
  maturity level of student, 204
  meaning and relevance, 204
  new procedures and methods, 211–212
  orderly sequence, 204
  planning activities, 203–204
  primary study extended, 204
  principle or concept, 204
  purposes and goals, 204
  safety, 215–218
  scientific methods, 204
  significance of data, 204
  simplicity of design, 204
  situation concerning life, 204
  solution not in textbook, 204
  staining procedure, 213
  student involvement, 207–210
  student laboratory assistants, 208–210
  student on his own, 204
  time and labor-saving, 214–215
    care and operating instructions, 214
    color coded slides, 214
    desk blotter-size calendar, 214
    files, 214, 215
    ordering supplies, 215
    schedule, 214
    supply catalogs, 214
    Want list, 214
  work centers, 212

Labor-saving in lab, 214–215
Lab report, 46
Lab session, 46
Leaflets, 215
Learning:
  mastery, 66, 70, 85
  rates, 47, 85
  reinforcement, 96
  stimulate, 43–63 (see also Teaching techniques)
Learning unit, writing, 69
Lesson guide, 55
Library resources, 34–36, 45, 137–140
Life:
  beating heart, 147
  determining when Sammy died, 147–149
  mercury amoeba, 146–147
  social/moral issue (see Social/moral issues)
  spontaneous generation, 142
Lifeboat Ethics, 158
LIFE Science Library Series, 60
Life situation, 204
Linear form, 67–68
Living laboratory, 23–24, 207–210
"Living on Lifeboat," 158
Living specimens, 71, 204–207, 212
Long-range assignments, 46
Long-range demonstration, 120
Lung capacity, 165–166
Lysozyme, bactericidal action, 111

M

Make-up work, 56
Malaria, cause, 142
Malpractice laws, 149
MANNO, 38
Marking papers, 189–192
Mastery learning, 66, 70, 85
Materials:
  illustrative, 70–72
  research, 137–140
  research projects, 100
Mathematics:
  blood counts, 125–126
  Chi Square test, 129–131
  DNA coding for amino acids, 123

Mathematics (cont.)
    doubling of cells in cell division, 122–123
    genetics studies, probability, 126–129
    population densities, 123–125
    standard deviation, 129
    tool of biologist, 131
Maturity level, 204
Measurement:
    microscope, 162–164
    turbidity, 164
Mercury amoeba, 113–114, 146–147
Metabolic rate determination in small organisms, 119
Metabolism, 171–174
Methylene Blue Reductase Test, 171
Microbiology, minicourse, 80–81
Microscope:
    aligning stage and ocular micrometers, 163
    depth or thickness of specimen, 164
    "microscopic" term, 163
    ocular micrometer, 162–164
    viewing and measuring, 162
Microscope slides, 71
Miller method, 119
Minicourses:
    benefits, 81–82
    full semester course, 81–82
    microbiology, 80–81
    starting, 80–81
Mnemonic devices, 36
Models, 26
Molecular motion, 211
Motility of microorganisms, 211
Motivation:
    atmosphere, 18–22
    games, 37–42
        board, preparation, 40–42
        evaluation, 39–40
        extend learning, 38–39
        review, 38
        stimulate learning, 37–38
    general reading, 34–36
    library resources, 34–36
    living laboratory, 23–24
    mnemonic devices, 36
    natural curiosity, 22–26
    outside speakers, 31–32
    relevance, 26–28

Motivation (cont.)
    research projects, 100
    research-type activity, 22
    slow learner, 96
    student interests, 22–26
    student involvement, 28–31
        decision making, 29–30
        evaluation, 30–31
        planning and preparation, 29
    students help one another, 32–33
    supplement the text, 34–36
    treasure hunt, 24–25
    turn students off, 17–18
    turn students on, 17–18
    visits, 33–34

N

New topic, introducing, 52–54, 66
Nigrosin stain (10%), 211

O

Oak Tree Basic Biology in Colour Series, 60
Objectives:
    course, 43
    learning, 54–56
Ocular micrometer, 162–164
"One Puff of Death," 155
One-to-one encounters, 181
On-going demonstrations, 120
Open book and/or notebook tests, 181
Open classroom, 101–102
Open-ended laboratory investigations, 44, 66
Opinion, expressions, 56
Opposition to Vaccination for Smallpox, 142
Optical density, 164, 165
Options, individualized study, 85–104 (see also Individualized study options)
Oral tests, 182–183
Ordering equipment and supplies, 215
Organization, data, 56
Origin of life, 157
Oscilloscope, 176
Osmosis, 115
O_2, photosynthesis, 110

Outside speakers, 31–32
Outside work, credit, 100–101
Overweight, 167
Oxford/Carolina Readers, 47, 61, 79
Oxidative enzyme, fresh produce, 110
Oxygen breathed by fish, 109

P

Paperback biology unit books, 61
Penicillin, 133–134
Personal assessment inventories, 193–195
Personalized learning program:
   assignments, 89–90
   Biology Book of Records, 90–92
   developing a scenario, 87–89
Petri dishes, sterile, 211
Photosynthesis:
   game, 42
   O_2 production, 110
   tracing pathway of carbon, 175
Phototropic response, euglena, 112
Pictures, 71
Placing an issue on trial, 151–152
Planning, 29
Plant Identification, 37
Plant tissues, conduction, 115
Plastic embedded specimens, 71
Pneumograph, 168
Point scoring system, 191
Poisons, 172
Pollution, water, 174–175
Population, 152–154
Population densities, 123–125
Population genetics, 42
Power's method, 119
Practice tests, 193
Predictions, skills for making, 144–146
Predictive form, 68
Prentice-Hall Foundations of Modern Biology Series, 60
Preparation, 29
Pretests, 180–181
Probability in genetics studies, 126–129
Programmed learning:
   active responses, 66, 72
   autotutorial, 66, 68–70, 85 (see also Autotutorial program)
   branching form, 68

Programmed learning (cont.)
   contractual basis, 73
   evaluation frames, 193
   facilities, 68
   goals and objectives, 72
   illustrative materials, 70–72
   immediate feedback, 66
   independent study projects, 66
   individual conferences, 73
   linear forms, 67–68
   logical, orderly sequence, 66
   mastery, 66
   new topic, 66
   open classroom/laboratory atmosphere, 73
   open-ended investigations, 66, 72
   options, 72
   outcomes, 68
   pitfalls, avoiding, 69–70
   predictive form, 68
   repeating program, 66
   reviewing topic, 66
   self-pacing, 67, 72
   slow learner, 66
   small group discussions, 73
   successful use, 66
   supplementary material, 66
   value sessions, 73
   varying its use, 66
   writing unit, 69
Progress of student, evaluating (see Evaluation)
Projecting, demonstrations for, 113–115
Pronunciation, 53
Protein digesting action of bromelin, 113

Q

Questions:
   responsibility for statements, 57
   students asking each other, 58
Quinlan case, 149
Quizzes:
   daily, 180
   teacher-made, 181

R

Rates of learning, 47, 85
Readiness, research projects, 100

Reading, general, 34–36
Reading materials, 71
Reference center:
adjunct, 61
assistance in selections, 61
benefits evident, 61
Biological Abstracts, 59
"borrowing" records, 60
carrel or table, 60
convenience and guidance, 61
current topic material, 60
facilities, 60
free or unscheduled time, 60
interest levels, 60
laboratory investigation, 61
match materials to students, 60
monographs, 60
over-night materials, 60
over-night rule, 60
pamphlets, 60
preferences, 61
privacy for learner, 60
programmed learning series unitexts, 60
reading ability levels, 60
responsibilities of student, 61
series booklets, 60
single topic books, 60
wide array of materials, 59
Reinforcement of learning, 96
Relevance, 26–28
Religious beliefs, 149
Reports:
format, 43
individual, 180
individual lab, 180
unsatisfactory work, 199
written, 61–63
Reproduction, human, 156–157
Researchers, 157
Research materials, 137–140
Research paper, 46
Research projects, 100
Research report analysis, 145
Research-type activity, 22
Resources:
library, 34–36
library and classroom, 45
research projects, 100
Respiratory health, 165
Respiratory rate, 172

Respirometer, 173
Responses, active, 66, 72
Responsibility:
statements, 57
student in reference center, 61
Review lessons, 193
Rights of individuals, 149
Right-to-Die, 149
Row-Peterson Unitexts, 60

S

Safety, laboratory, 43
Safety in lab, 215–218
Scale, effect on organisms, 134–135
Schatz, Vivian and Albert, 155
Schedule, 214
Science fiction writers, 157
Scientific American offprints, 47, 60, 79
Scientific attitudes, 142–146 (*see also*
Attitudes)
Scientific methods, 204
Sea urchin, 212
Self-evaluation, student, 192–195
Self-pacing, 67, 72
Sequence:
investigation, 204
programmed learning, 66
Serial numbers, 214
Sewage, 174
Short answer test, 182
Shot-gun fungus, 212
Silicone culture gum, 211
Skeletal muscle fibers, contraction, 115
Slides:
color coded, 214
selected, prepared, 71
35 mm., 187
Slow learner:
individualizing learning, 95–97
programmed instruction, 66
Smallness, metabolic rate, 172
Smallpox, 142
Smoking, 154–156
Social/moral issues:
cloning, 157
Evolution-Creation, 157
genetic engineering, 157
human reproduction, 156–157

INDEX 251

Social/moral issues (cont.)
  life and death, 149
  long-range philosophies, 157
  old and new, 157
  origin of life, 157
  population, 152–154
  predictions of researchers, 157
  science fiction writers, 157
  smoking, 154–156
  techniques, 150–152
    debates, 150–151
    discussions, 151
    individual enrichment experiences, 152
    interviews, 150
    placing issue on trial, 151–152
  theories, 157
  value judgments, 149
Socio-biological topics, 136–137
Sodium hypochlorite, 211
Speakers, 31–32
Specimens, 71, 204–207, 212
Spectrophotometer, 164–165
Spell-Down, 37
Sphygmomanometer, 166–167
Spirometer, 165–166
Spontaneous Generation of Life, 142
Staining procedure, 213
Standard deviation, 129
Standardized tests, 181
STERIFIL apparatus, 174
Strokes, 167
Student involvement, 28–31
Student laboratory assistants, 208–210
Students, helping one another, 32–33
Study, how to, 46
Study habits, 44
Style, teaching, 47–48
Supplementary materials, 44
Supplements to text, 34–36
Systolic Pressure, 167

T

Taping, demonstrations for, 113–115
Teacher-made tests and quizzes, 181
Teaching style, 47–48
Teaching techniques:
  biology as a process, 51

Teaching techniques (cont.)
  biology student's handbook, 43–44 (see also Handbook)
  change of pace, 51–52
  classroom reference center, 59–61
  develop a style, 47–48
  group interaction, 56–59 (see also Group interaction)
  Guidelines for Biology Students, 44–46 (see also Guidelines for Biology Students)
  individualized learning, 47
  innovation, 65–83 (see also Innovation)
  interest, high level, 50–51
  introducing topic, 52–54
    classification, 53
    dramatic anecdote, 52–53
    groups of related words, 53
    issues, 52–53
    naming structures and processes, 53
    new terms, 53–54
    proper pronunciation, 53
    student analysis of words, 53
    verbalizing terms, 53
    word recognition and usage, 53
    words-in-series, 53
  objectives of lesson, 54–56
    agenda for day, 54
    assignment, 54
    data-chart association, 56
    designing own charts, 56
    evaluation, 54
    films or film strips, 56
    guide sheets, 54, 56
    make-up work, 56
    organizing data, 56
    sample lesson guide, 55
    work sheet, 56, 57
  variety and balance, 48–50
  welcoming students, 44–46
  written reports, 61–63
Techniques and instruments, 161–177 (see also Instruments and techniques)
Terms, new, 53–54
Tests (see Evaluation)
Text, supplements, 34–36
Theories, 157
35 mm. slides, 187
Thunberg technique, 170–171
Time, research projects, 100

Time distribution, topics studied, 43
Time-lapse demonstrations, 111–113 (see also Demonstrations)
Time-saving in lab, 214–215
Topics:
  biochemical-biophysical, 135–136
  new, 52–54, 66
  outline, 43
  reviewing, 66
  socio-biological, 136–137
  time distribution, 43
Transpiration, plant leaves, 109–110
Treasure Hunt, 24–25
Trial, issue on, 151–152
Turbidity, 164
T.V., closed circuit, 108–109

U

Unit, writing, 69
Unit tests, 187–189
U.V. light, sensitivity of bacteria, 111–112

V

Vaccination, smallpox, 142

Variety, 48–50
Verbal communications to students, 44
Viewing, microscope, 162–164
Visits, 33–34
Vital Capacity, 165, 166
Volunteers, student, 118–119

W

Want list, 214
Warranties, 214
Water pollution, 174–175
Winkler method, 119, 169
Word games, 38
Word hunts, 38
Words, new, 53–54
Work centers in laboratory, 212
Working models, 26
Work sheet, 56, 57
Written reports, 43, 61–63
Written tests, 182–183

Z

ZOOLO, 38